T0317434

HORIZON WORK

Horizon Work

At the Edges of Knowledge in an Age of Runaway Climate Change

Adriana Petryna

PRINCETON UNIVERSITY PRESS

PRINCETON AND OXFORD

Published by Princeton University Press
41 William Street, Princeton, New Jersey 08540
6 Oxford Street, Woodstock, Oxfordshire OX20 1TR

press.princeton.edu

The quotation on p. 86 by Robin Wall Kimmerer is from *Braiding Sweetgrass* (Minneapolis: Milkweed Editions, 2013). Copyright © 2013 by Robin Wall Kimmerer. Reprinted with permission from Milkweed Editions. milkweed.org

All Rights Reserved

Library of Congress Cataloging-in-Publication Data
Names: Petryna, Adriana, 1966– author.
Title: Horizon work : at the edges of knowledge in an age
   of runaway climate change / Adriana Petryna.
Description: Princeton : Princeton University Press, [2022] |
   Includes bibliographical references and index.
Identifiers: LCCN 2021026124 (print) | LCCN 2021026125 (ebook) |
   ISBN 9780691211664 (hardcover ; acid-free paper) | ISBN 9780691232591 (ebook)
Subjects: LCSH: Climate change mitigation. | Climatic changes—Forecasting. |
   Climatic changes—Social aspects. | BISAC: SOCIAL SCIENCE /
   Anthropology / General | SOCIAL SCIENCE / Technology Studies
Classification: LCC TD171.75 .P38 2022 (print) | LCC TD171.75 (ebook) |
   DDC 363.738/746—dc23
LC record available at https://lccn.loc.gov/2021026124
LC ebook record available at https://lccn.loc.gov/2021026125

British Library Cataloging-in-Publication Data is available

Editorial: Fred Appel and James Collier
Production Editorial: Lauren Lepow
Jacket Design: Layla MacRory
Production: Erin Suydam
Publicity: Kate Hensley and Kathryn Stevens

This book has been composed in Adobe Text and Gotham

Printed on acid-free paper. ∞

Printed in the United States of America

10 9 8 7 6 5 4 3 2 1

For João, Andre, Tania, and Noemia

CONTENTS

ILLUSTRATIONS

HORIZON WORK

Prologue

Epic storms from warming oceans, rising sea levels, extreme heat, prolonged droughts, catastrophic wildfires—the cool directness of the steeply climbing line of carbon dioxide emissions fails to match the palpable sense of environmental crisis those emissions provoke. And it's not just the physical climate that is changing: our expectations for how the environment should act are being constantly shattered. Some still prefer not to acknowledge this increasing divergence between expectation and reality; the culture of climate change denial ignores the problem full stop. Others embrace doomsaying in order to catalyze action through fear, while still others worry that too much doomsaying will lead to hopelessness and inaction.

Meanwhile, a familiar image of nature as stable is now gone. Rapidly departing from fairly predictable patterns and historical trends, nature itself has entered a runaway state. This is especially true in the United States, where a long-standing focus on the suppression of wildfires, aiming to turn them into

Atmospheric CO$_2$ concentration
Global average long-term atmospheric concentration of CO$_2$ (parts per million)

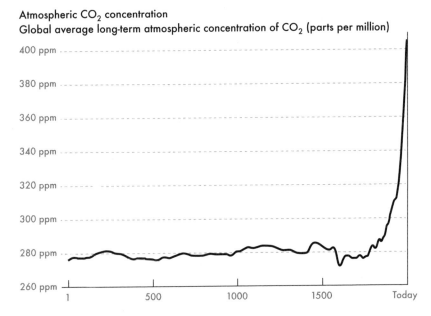

FIGURE P.1. Rising atmospheric concentration of CO$_2$ (https://ourworldindata.org
/co2-and-other-greenhouse-gas-emissions).

something humans can banish, has also fueled their likelihood.
As more and more people are either involved in "fighting" to
contain fires or displaced by them, how we respond to emergencies to slow the pace of destruction and how we organize
emergency response in the first place become open questions.

This book focuses on those questions by considering how
different communities of experts, including climate and wildfire scientists, emergency managers, first-line responders, and
Indigenous knowledge holders, reckon with breakaway ecological processes that deny a coherent vision of control, leaving them staring at the edge of what we can see and know.
Yet as they face this edge of knowledge, these experts, rather
than resigning themselves to either hopelessness or despair,

are creatively looking for new options, transformations, and outcomes. How do they do this? And what can the rest of us learn from them?

As one wildfire scientist at the Rocky Mountain Research Station put the problem to me in 2018, "We need to acquire a horizon." As our expectations shatter and models for handling crises are outpaced by what is happening, the present becomes increasingly defined by urgency. In such a scenario, a horizon becomes a tool—a kind of lever to pry time away from the pace of runaway climate change in order to make room, as the scientist put it, "for deciding what we actually are going to do over the next week, not like the next hour." In this context, the expansion of a deliberative space and time amid onrushing disasters becomes crucial work. This work, which I call "horizon work," allows experts and the public to find other meaningful points of reference from which to imagine how to organize a response to the current crises, before we lose the capacity to respond.

In the following pages, conversations with different thinkers, observers, and eyewitnesses show the breadth of horizon work. It includes rethinking how to see our world and interpret and respond to its shifts; how scientific research and management paradigms negotiate the decreasing reliability of models and projections; and how emergency response systems contend with increasingly destructive climatic changes that put more lives and communities at risk. How these different forms of "work" relate to each other and bolster each other, without marginalizing one another, points to political transformations that are necessary in configuring new, livable horizons.

A part of this inquiry will take us to ecological theorists and experimentalists who, over the last few decades, have reckoned with large-scale ecosystems and are attempting to define varieties of thresholds (so-called tipping points) that, if crossed, may entail irreversible shifts. These shifts can appear at first as anomalous one-offs, but later reveal themselves to be the new normal. Take, for example, mountains that should have been soaked with rainwater or covered in snow, and thus resistant to fire. However, within a stretch of a few years, supposedly unburnable peaks succumb to devastating wildfires that occur outside of a usual fire season. Ecological regime shifts can be stealthy and all the more dangerous because their warning signals are sometimes obscure or undetectable. The picture we get is partial: we understand some aspects of the change but not others, and, sometimes, the realization that a shift is irreversible comes much too late.

Amplified uncertainty around how to anticipate rapid environmental change impedes emergency response protocols, particularly to wildfires. In talking with wildfire scientists, foresters, and frontline emergency workers in the United States, I learned about what human, scientific, and ethical aspects of decision making look like under extreme uncertainty, and where models and expectations are not able to keep up with the frequency and severity of today's emergencies. As ecological shifts collide with existing paradigms of response, the climate crisis powers this collision into the future.

But if all we see in this instability is doom, we will have missed something crucial about how we are implicated in enabling this collision, and therefore face a choice. We can continue with the status quo, holding to a myth of stabilizing nature at all costs that still holds sway. Or, in the words of one fire scientist, we can "act in a way that we are not cutting off options for future generations." As we embrace the latter, necessary choice, we are challenged to think differently, to see processes that weren't thought possible. In helping to configure this shift,

I use a rhetorical device, anthimeria, which turns words into novel grammatical forms. For example, when Shakespeare writes, "And I come coffin'd home," he abolishes the division between noun and verb; Milton's "palpable obscure" and "the vast abrupt" do the same between adjective and noun. Both linguistic shifts make room for unexpected meanings within established parts of speech.

In the same spirit, I turn the noun "horizon" into a verb form, *horizoning*, and use it as a conceptual device for thinking about and responding to complex futures. Along the way, I hope to discover new range and even circumstances for action that otherwise seem precluded by the disastrous onrush of runaway change. When the horizon is considered in the verb form, harrowing scenes of wildfire are not inevitable; rather, there is room for distinctions to be made between inevitability and choice. I take my cue in this from the climate and wildfire scientists and the emergency responders I came to know in the course of this inquiry. They were less interested in the question of how far, close, or beyond we are with respect to abstract tipping points. Rather, where projections falter, horizon work begins.

1

What Is the Upper Limit?

The breath you just took contains over 400 parts of carbon dioxide per million molecules (ppm) of air.[1] People living at the start of the Industrial Revolution would have inhaled about 278 ppm. Since then, levels of CO_2—the leading greenhouse gas driving changes in the climate—are on course to double owing to the relentless burning of fossil fuels. In a worst-case scenario, CO_2 concentrations will exceed 900 ppm by the year 2100. Unfortunately, that scenario is within the realm of possibility. Carbon dioxide is the natural product of cellular respiration in animals and plants. Fossil fuel emissions from human activity over the past two centuries now threaten our atmosphere, oceans, and life on Earth. In spite of the impacts—extreme heat and wildfires, catastrophic floods and storms, massive crop failures, and unrelenting biodiversity loss—some experts have made the claim that human cognition operates on a very narrow spatiotemporal scale; we are unable to see—let alone deal with—the flood of changes that we have unleashed. Our horizons are so limited, the argument goes, because *Homo sapiens* never evolved enough mental bandwidth to apprehend a long-term future. Our ancestral selves were mainly preoccupied with the "immediate band, immediate dangers, exploitable resources, and the present time."[2] So here we are, built to be blindsided in a new and hostile world. Yet the claim of cognitive barriers is just that—a claim—and, in any case, overcoming such barriers to responding to all but our short-term needs is not the real challenge. Rather, we need to ask how narrowed self-understandings prevent us from effectively addressing the problem of climate change, leaving us stranded in a present that may not be survivable.

More than a century's worth of research undercuts the idea that a bias toward inaction in a high-CO_2 world is preordained. During World War I, when submarines were first widely deployed in warfare, a US Navy sanitary officer and surgeon named R. C. Holcomb worried about carbon dioxide displacing oxygen in breathable air in these sealed underwater capsules. Carbon dioxide is a colorless and odorless gas, so it is tempting to think that its risks cannot be sensed. Holcomb questioned this assumption, writing, "We cannot forget that we are at the bottom of an aerial ocean and saturated with its gases." He expressed concerns over "men obliged to breathe their own expired air over and over again."[3] More than a hundred years later, we think of carbon dioxide in more distant (atmospheric) terms, an input to be tracked or mitigated in climate change scenarios. Its physiological impacts are harder to grasp. Holcomb made his observations at a time when, in military and medical spheres, new instruments were being devised that could scrub carbon dioxide from closed environments. Consider the American pharmacologist Dennis Jackson, who wanted to make anesthesia gas accessible to his poorer surgical patients. Breathing chambers of the early twentieth century delivered

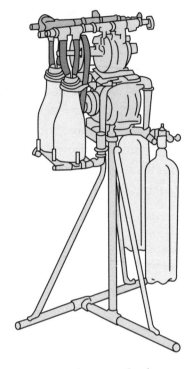

FIGURE 1.1. Jackson CO_2 Absorber (redrawn from image courtesy of Wood Library Museum).

expensive nitrous oxide, but they also leaked it. Hoping to make its delivery more efficient, in 1914 Jackson invented a closed circuit chamber to trap the nitrous oxide. But it also trapped patients' exhaled carbon dioxide gas. When he added soda lime, which absorbed the gas, patients could rebreathe expired air. It so happened that the "Jackson CO_2 Absorber" was invented in St. Louis, a city once saturated with coal smoke. The absorber worked so well that when Jackson tested it on himself, he reported having "the first breaths of absolutely fresh air he had ever enjoyed in that city."[4]

Like atmospheres, our bodies require careful calibration between oxygen consumption and carbon dioxide production. The amounts of carbon dioxide that are present in our arterial blood and exhaled in our breath are always maintained reciprocally through a partial pressure gas exchange. This exchange is critical to survival. When the gas accumulates in our blood during sleep, our bodies signal an imbalance (by snoring, waking up, breathing abnormally deeply, or, if the lungs' ability to remove CO_2 is seriously impaired, exhibiting asthma or respiratory failure). Doctors use CO_2 saturation as a prognosticator for "time to death" in terminal patients.[5]

Too much CO_2 in the blood is a sure sign of imminent cardiac arrest or death.

So immediate are visceral responses to carbon dioxide overload that researchers have attributed to it involuntary reactions of all kinds. In work that was a precursor to his studies on "voodoo" death,[6] Walter B. Cannon, a professor of physiology at Harvard from 1906 to 1942, experimented on dogs to show how distress and panic increase the body's production of carbon dioxide, which he famously called the fight-or-flight response. "Great exertion, such as might attend flight or conflict," he wrote, "would result in an excessive production of carbon-dioxide."[7] More recently, researchers have found that they can simulate a variety of mental infirmities, from anxiety and panic disorders to combat-related stress reactions, by exposing human subjects to carbon dioxide–enriched air.[8]

Distress, an induced panic, or even cardiac arrest: our bodies respond to this insensible gas, whether we're conscious of its presence or not. Given the wide-ranging effects CO_2 has on biology, we can ask how much of a threat to physiological equilibrium we are willing to tolerate. In one respect, it is difficult to say: while the unconscious systems of our bodies are adept at signaling intolerance, the conscious ones are often too sluggish to recognize or fend off the danger.

Let's then move from the autonomic realm to the question of how awareness and assessment of CO_2's risks have evolved, drawing examples from modern agriculture and war. In 1954, when two Kansan farmworkers descended into a silo full of beans, barley, and oats, the gas released from the fermenting silage killed them. Silos notoriously contain high amounts of carbon dioxide, giving no warning of their lethality to people entering them.[9] So farmworkers developed homespun techniques to test for gas buildup before entering these structures.

One involved lowering a candle into a silo to see whether its flame died out (this occurs when carbon dioxide gas displaces oxygen needed for combustion). Another entailed suspending a warm-blooded animal in the structure to see whether it fell unconscious. When the sentinels' limp bodies were fished out of the silos, it was found that "an exposed guinea pig was unconscious within 30 seconds and a rabbit within 60 seconds."[10]

In an early study (1914) of a carbon dioxide accident on a farm, investigators found four men dead in a silo in Athens, Ohio. Coworkers reported that these men had entered the silo to tamp down new silage, but "within about five minutes the men inside were not responding to the shouts of their coworkers." Accident investigators noted CO_2's ability to trick the senses, writing that a "more peaceful and inviting scene could not be imagined than the warm, pleasant smelling green silage within."[11] Sensory trickery of this kind also has its uses: for decades, farm managers have been exposing livestock to high levels of carbon dioxide to anesthetize them before slaughter, a method that animal welfare advocates consider more humane than electrical stunning.[12]

As examples from agriculture illustrate, knowledge of the effects of carbon dioxide is carved into modern life. That humans can do no more than deny them because we as a species cannot see past our arms does not add up. History, too, refutes this notion. When incendiary bombs were dropped during World War II, European cities were flooded with clouds of toxic gas (including CO and CO_2), killing untold numbers of people for whom overcrowded air-raid shelters provided no escape.[13] In July 1943, the air raids on Hamburg ignited massive fires. The author of *The Night Hamburg Died* (1960) describes what transpired in the shelters from these torrents: "Sealed into their cellars, huddling behind heavy doors, they have closed themselves off from the outer world and the oceans of fire splashing around and over their warrens. No flame ever touches them, but not a man, woman, or child survives. Not a single living soul. Not a human being, an animal, not even the smallest rodent, not a single insect, survives."[14]

There was also neither warning nor escape when, on August 21, 1986, an underground bubble of carbon dioxide erupted in Lake Nyos, an active crater lake in Cameroon, releasing a low-hanging gas cloud that killed over seventeen hundred people.[15] One survivor, knocked unconscious for several hours, described his experience when he woke up: "I could not speak . . . I could not open my mouth because then I smelled something terrible . . . I heard my daughter snoring in a terrible way, very abnormal." He continued: "When crossing to my daughter's bed . . . I collapsed and fell . . . My daughter was already dead . . . I got my motorcycle . . . As I rode . . . I didn't see any sign of any living thing."[16]

An American biologist who studied the Lake Nyos disaster (and another at Lake Monoun in Cameroon two years later) conveyed to me some of the physical and sensorial aspects of total exposure: "At the heart of the cloud released during the Lake Nyos and Lake Monoun disasters, the concentration of CO_2 was 100%—that is, the CO_2 had displaced all of the normal air that we breathe." Concentrations of CO_2 above 15–20 percent will cause suffocation and death in animals and humans.[17] In a lower range of 10–15 percent, delusions can set in. Here, as the scientist described to me, "CO_2 can act as a sensory hallucinogen, such that people feel and smell things that aren't really there." Where the CO_2 concentration hovered just below the lethal limit, some Lake Nyos survivors reported smelling rotten eggs or gunpowder and feeling very warm. "The rotten eggs smell is unmistakably a smell of sulfur gases and feeling warm is also associated with volcanoes producing heat," he noted. "However, our analyses showed that there were no sulfur gases released (or very little) during the disaster, and that the gas burst was not associated with heat release from a volcano."[18]

In other words, the gas cloud the biologist describes was full of sensory bewilderments, resulting from a freak geophysical event the likes of which most of us will never experience. But I knew someone who may have lived through something comparable. My father was a twelve-year-old child refugee from a small village in Ukraine—one among hundreds of thousands who fled the country for displaced persons camps in Western Europe when the Soviet and German forces met in 1944. Allied forces conducted aerial bombing raids, targeting industrial plants and railway stations as well as fleeing civilians, as he would point out. The civilian refugees were a hundred miles into their trek when one of the bombs from a shuttle bombing operation fell near a border town, hitting an underground tunnel that served as a makeshift bomb shelter. His older sisters had not made it to the overcrowded shelter-turned-death-pit—but he had. Through a child's eyes, he described to me what it was like to be packed inside and, in his words, "what people's lungs look like when they are gasping for breath." By some miracle, the little boy found himself near a tiny airhole. Taking in small sips of fresh air, he observed the terrifying distensions all around him. He lost consciousness and, along with other presumed-dead bodies, his was thrown onto a flatbed truck. The high-pitched voice of his oldest sister calling out his name (Misio!) awoke him, and then (a detail that as a child I could hardly fathom) he stood up from the pile of bodies and got off the truck. The small amount of oxygen from that hole in the tunnel prevented the extreme CO_2 concentrations from killing him.

This near-fatality conjoins histories of human breath and pyrogeographies of modern warfare. In his essay "Air War and Literature," the writer W. G. Sebald depicts the absolute destruction wrought by the Allies' aerial bombing of European cities in World War II. There was a narrative vacuum. German writers, Sebald argued, "would not or could not describe the destruction of the German cities as millions experienced it." The bombings left "31.1 cubic meters of rubble for every person in Cologne and 42.8 cubic meters of rubble for every inhabitant of Dresden."[19] Adding to the physical destruction, the Hamburg air raids produced a massive urban firestorm, five kilometers in height and covering seventeen square miles.[20] Winds produced a high-velocity fire whirl that still perplexes fire scientists today. Of Hamburg's obliteration by fire, Sebald wrote: "At one twenty a.m., a firestorm of an intensity that no one would ever before have thought possible arose. . . . At its height, the storm lifted gables and roofs from buildings, flung rafters and . . . billboards through the air, tore trees from the ground and drove human beings before it like living torches."[21] Scenes like these, along with unrecognizable ecological synergies, are at the heart of these overlooked embodiments of total war.

An estimated forty-five thousand died in the aerial bombings. Their incendiary effects, along with those of nuclear weapons, led to an "unprecedented boom in the research of wildland fires."[22] But the boom was short-lived. In the 1950s and 1960s, when Cold War researchers were conceiving of radioactive fallout shelters to protect people in the wake of nuclear attacks, they overlooked the fact that shelters would ultimately be "useless, largely because of firestorms."[23] They narrowed the scope of the hazard to a mechanical balancing of oxygen supply with carbon dioxide removal in closed environments. How long could occupants live in a nuclear fallout shelter? Studies tested chemical

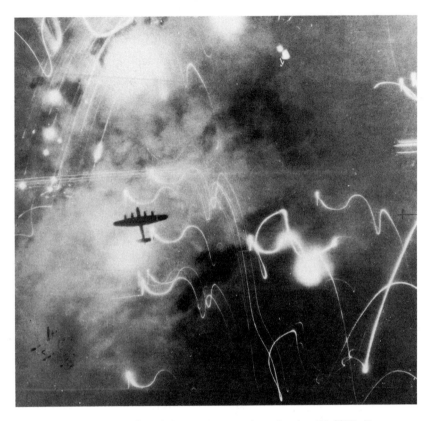

FIGURE 1.2. Bombing of Hamburg. Avro Lancaster heavy bomber, World War II, 1939–1945 (Science & Society Picture Library).

carbon dioxide removal as a method of prolonging occupancy after breathable air dissipated. In one study, two adults sat in a fallout shelter as researchers monitored oxygen consumption and carbon dioxide accumulation.[24] In hour one of occupancy, the oxygen remained at 20 percent. In hours two and three, it dropped to 19 percent. In hour four, it was at 18.5 percent. Carbon dioxide concentrations rose steadily, from 0.5 percent in hour one to 1.7 percent in hour four. In a bomb shelter packed with hundreds of people, this rate of increase would likely result in CO_2 gas concentrations in the range of 10 percent, if not more, certainly high enough to cause them to fall unconscious or die.

As with any other noxious gas, carbon dioxide is classified as an occupational hazard; its levels are monitored and federally regulated in various industrial settings to insure safe respiration. The US Department of Labor, for example, considers 400 ppm to be the outdoor norm for CO_2 exposure, and 800 ppm the indoor norm. According to a CO_2 monitor salesman I spoke with, 1,500 ppm "is when you start to see effects." In fact, the majority of his sales were to school districts because of concerns about the dangers of carbon dioxide to children's school performance: "We need to break up the CO_2 concentration in schools." At 5,000 ppm, metabolic stress and narcosis or a depressed state of consciousness can set in.

Seen through its somatic history, carbon dioxide comes to be palpable through industrial techniques and standards developed to exploit its potentials, mitigate its harms, or protect breath. That history consigns humans and nonhumans (rodents, cattle, and refugees) to the structures of research and the rubble of modern war. It also becomes an exercise in securing what the philosopher Achille Mbembe calls "the universal right to breath." Following the death of George Floyd, whose public assassination by police chokehold ignited protests against racist policing and anti-Blackness around the world, Mbembe writes, "Caught in the stranglehold of injustice and inequality, much of humanity is threatened by a great [suffocation]" and this sense "spreads far and wide."[25]

Today, threats to breath are all around as "[w]e are adding planet-warming carbon dioxide to the atmosphere at a rate faster than at any point in human history since the beginning of industrialization."[26] CO_2 toxicity has been calculated extensively (from the science of the fight-or-flight response to occupational safety and even bomb shelter survival). When it comes to planetary risk, a terrible disjuncture remains between the scale of the threat and the pace of collective efforts to stop its cascading impacts. There is a failure of imagination, which the writer Amitav Ghosh calls a "great derangement," when it comes to connecting the burning of fossil fuels and CO_2 rise to our altered present. Politicians with no vision beyond the next election cycle normalize the derangement, or the idea that our horizons, so truncated, will never allow us to meet conditions where they are.

Meanwhile, as we will see in this book, earth scientists are getting a better handle on how increases in CO_2 and other fossil fuel emissions threaten to destabilize entire Earth systems. Having passed a particular threshold, ocean acidification—caused by the overabundance of CO_2 in the seas—will trigger widespread fish extinctions due to diminishing coral reef ecosystems (which sustain roughly 10 percent of the world's fisheries). On land, rising temperatures associated with increasing CO_2 concentrations threaten to wipe out agricultural production in some areas.[27]

Carbon dioxide is absorbed in the atmosphere and by forests and oceans. But what kinds of worlds will be habitable once parts of the Earth system have lost their ability to "scrub" carbon dioxide? Researchers are unsure about where the CO_2 will go. The future of Earth's CO_2-offsetting reservoirs (or carbon sinks) is uncertain—nearly a third of them are saturated or have disappeared. This occurs at a time when CO_2 levels routinely exceed 400 ppm, higher than they've been since "three to five million years ago—before modern humans existed."[28]

I measured levels of the gas in my everyday (pre-COVID) surroundings with a handheld CO_2 monitor that I purchased online. There was a surprising amount of variability. The CO_2

in my small office measured 608 ppm; a lecture hall, 955 ppm; a room where I met with a group of incoming college students, 1,027 ppm. When I stuck the monitor outside my office window, it read 388 ppm. At home, levels varied from 402 ppm to 1,339 ppm. When I exhaled right into the monitor, it jumped to 3,994 ppm. Variability, I learned, is the very thing that has allowed land animals to survive in milieus with relatively high levels of CO_2—and humans to dominate the planet. If the CO_2 is too high in one setting—say, in a classroom or office—we will know it (perhaps not consciously) and eventually leave the room or open a window for fresh air. Even if we start hyperventilating, we can usually recover, which, strictly speaking, means returning our partial pressure of carbon dioxide (a measure of carbon dioxide in arterial blood) to a normal level.

As air-breathers, humans have a high partial pressure of carbon dioxide (PCO_2). Our bodies are equipped to deal with variable CO_2 levels. In the constant adjustment to variability, we normally have the luxury of forgetting that without such adjustment, we would soon be dead. Contrast this with aquatic animals, for whom "the difference in PCO_2 between inspired and expired medium," in this case, water, is much smaller.[29] The smallest rise in CO_2 in any aquatic system can trigger a state called hypercapnia, from the Greek *hyper* (over) and *kapnos* (smoke) and occasion a massive fish die-off. Aside from the very few fish that can air-breathe (using their mouths, esophagi, or stomachs to trap air when water becomes oxygen-deprived), water-dwellers, for the most part, can't compensate for variability in their aquatic environments the way that air-breathers can, nor can they escape water in which they cannot breathe. Readers may have seen the workings of hypercapnia in oxygen-depleted ponds or lakes. One day, everything seems normal, as life teems just beneath the surface; the next day, fish underbellies cover the entire lake as far as the eye can see.

FIGURE 1.3. Lake ecosystem regime shift after human pollutants decrease oxygen levels, Rio de Janeiro, 2013 (Reuters/ Alamy/ Sergio Moraes).

We may find comfort in the fact that we are not fish. Air is a much more forgiving medium than water as far as respiratory physiology goes. But when it comes to humans and fish, how should we conceptualize differences in survival capacities amid elevated CO_2 levels? Is it a matter of physiological difference (that confers some seemingly inherent advantage in one kind of animal and not another)? Or is it a matter of an environmental difference (that will always provide one kind of animal and not another with escape hatches within variable milieus)? Setting species-specific distinctions aside, is there a place and time in which human and fish fates might converge, pushing us toward some edge, some horizon beyond which existence ceases to be viable—call it extinction—without our even noticing?

2

Building Perceptual Range

Sudden and irreversible ecological shifts have perplexed scientists for more than a century. American ecologist Stephen Forbes, a founder of aquatic ecosystem science, studied high fish mortality in Lake Mendota in Wisconsin. In his 1887 essay "The Lake as a Microcosm," Forbes wrote, "The animals of such a body of water are, as a whole, remarkably isolated—closely related among themselves in all their interests, but so far independent of the land about them that if every terrestrial animal were suddenly annihilated, it would doubtless be long before the general multitude of the inhabitants of the lake *would feel the effects of this event in any important way.*"[1]

Some animals exhibit a sixth sense toward impending disaster. The goats and sheep that live on Mount Etna in Sicily reportedly move to safe areas or wake up at night at the onset of a volcanic eruption. Dogs in Tohoku, Japan, were able to detect an earthquake's seismic activity before humans did.[2] The expression "the calm before the storm" refers to a quiet gathering of senses before a period of turbulence. Forbes's scene also

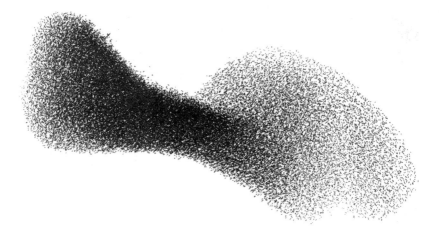

FIGURE 2.1. European starling flock (Manuel Presti/Science Source).

evokes a calm, but one that is coupled with a fatal nonfeeling. The fish perish from sensory myopia, surrendering to the insensible and inevitable.

I take this scene as a starting point for considering environmental instability and its links to anticipatory thinking.[3] Can humans shed myopia for the sake of survival? The "ancient brains" we have supposedly inherited become part of a Forbesian mental furniture whereby we cannot see beyond a certain horizon and thus coordinate for long-term survival. Such fatal belatedness may afflict humans more than some other animal species. Ants, for example, rely on neighbors, chemical cues, and other signs to coordinate colony behavior.[4] Birds rely on mutual proximity to hedge against an existential threat, synchronizing individuals' behavior to confuse or escape predators.

Yet when it comes to climate change, even the ability to act cooperatively may not be enough to allow creatures to survive.[5] For species pushed beyond the envelope of their adaptive capacity, options are limited. They can include rapid adaptation (a culling process almost always involving high mortality), migration to more suitable ecological niches, or extinction. As climate conditions change, and as the range of suitable habitable zones for a species "may shrink to nothing," options one and two may no longer be available.[6]

What remains is an extinction process that, in the first instance, involves growing insecurity over whose envelope will shrink next. In their struggles to adapt, interactions between predator and prey can fall out of sync. Climate-driven geographic range shifts affect competition for food, but the pace of ecological actors' adaptations can vary. Increasing temperatures can affect one animal differently from another. A predator reaching a threshold of heat intolerance moves up a mountain to find cooler temperatures. This move can create a mismatch in predator-prey interactions, as the prey may not follow at the same pace.[7] Climate-driven range shifts that affect food competition may strand some species (in this case, a predator) in the wrong climate envelope.

What began as a careful choreography of survival now entraps actors in their demise. By the time this entrapment sets in, all kinds of extinction scenarios have started to play themselves out. Witness the image of emaciated sea lion pups, inadvertently abandoned by mothers desperately pursuing prey that were themselves forced out of their own habitable zones owing to warming ocean temperatures. Disjunctures between expectation and reality like these dissolve familiar rules of survival. All species are implicated in a muddle of extinction whose dimensions transcend knowable evolutionary pathways. There are other ways to conceive of this muddle, where signaling among human and nonhuman actors shapes efforts to defend the circumstances of life.

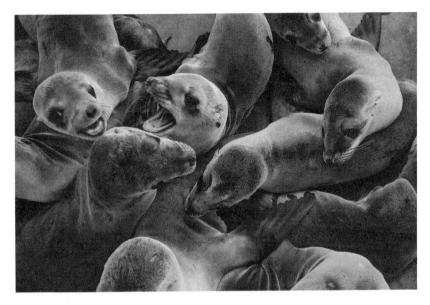

FIGURE 2.2. Rescued sea lion pups at the Pacific Marine Mammal Center in Laguna Beach, California (2015) (photograph © Kendrick Brinson; reproduced by permission of the artist).

Forbes projects a certain sensory myopia onto fish that makes them capitulate to an inevitable demise.[8] His meditation on aquatic microcosms relies on familiar tropes in which observers stand at a distance from a nature that is timeless ("unchanged from a remote geological period") or in a state of passive balance (involving a "general adjustment of numbers to the new conditions").[9] The aquatic ecosystem science founder made observations of fatal nonfeeling on the territories of the Ho-Chunk people, who in 1837 were forcibly removed to Iowa, Minnesota, and South Dakota, only to return to Wisconsin to live on the lands that the United States had stolen from them. Settler institutions and imperatives helped cement a "settler time" in which the reality of ongoing colonial and ecological violence was denied.[10]

A few years before Forbes's Lake Mendota study, there was a flood, and sewage and organic matter from nearby farms rushed into the lake. The perch were behaving oddly. Some began to feed in deeper waters, while others, according to one report, washed ashore by the hundreds of tons. Cartloads were taken away and buried, "but still the shore is covered with their carcasses."[11] The swallows and sparrows who ate the fish carcasses were dying too, with deaths so sudden that the carcasses had "no trace of decay." The fish carcasses were "in good average condition, often fat and plump," and seemingly healthy birds fell "without a flutter or any indication of pain."[12] Attendants at a nearby so-called State Insane Asylum, eager to fend off the prospect of a sky full of birds falling to the ground, caught a few and tossed them back into the air in the hope that they would resume flight.[13] Call these reactions absurd. But what if we the settlers are the asylum attendants, throwing up dead things that we think can still fly? Or the sea lion pups, caught up in the confusions of abrupt shifts and envelopes falling apart? As models of change will soon be missing a fundamental horizon of how life as an operative system stays intact, how is it that the loss of totalities, major ecosystems, refuses to take up residence in the imagination?

In an evolutionary universe infused with panic, the future dissolves into small moments of myopic surprise. One of Forbes's contemporaries, the philosopher Friedrich Nietzsche, echoed this sense of dissolution, warning that when a living thing has a too-rigidly-held or "hemmed in" horizon and is unable to "lose its own view" in another's, it "will come to an untimely end."[14] While different myopias lead to different kinds of ends, most nineteenth-century natural scientists hewed to a vision of evolution that subsumed untimely ends into a scheme of gradual change through natural selection. Filling in the blanks in a tale of minute variation through gradual differentiation was its scientific modus operandi, rather than poking holes in illusions that, in Nietzsche's words, we "shall have a still more splendid time."[15]

The gradualist version won out and precluded ideas of evolutionary jumping and abrupt change, which threatened "core British cultural principles of Darwin's day" and promised immunity from the fatal nonfeeling that plagued other species. As anthropologist David Napier writes: "dangerous biological transformation and upward advancement are, the English fortunately discovered, ruled by a safer gradualism" that precluded imagining worlds different from their own.[16]

Today, the rapid changes that are destabilizing Earth systems cannot be apprehended according to the model of safer gradualisms. Telling the time of extinction constitutes a new kind of natural experiment in radical uncertainty, in which scientists guess our "borrowed time" and develop measures of where different ecological actors are in a "final decline to extinction."[17] In this experiment, a few of those actors will thrive in new niches, but many will be stranded in old ones. Which living things will remain and which will disappear from the experiment is an open question.

No one, human or nonhuman, is excluded from this experiment. In 2020, the west coast of the United States saw its largest wildfires; in Brazil, a quarter of the world's largest tropical wetlands burned. The fires consumed Indigenous territories in both countries. Sandra Guató Silva, a community leader and healer of Baía dos Guató, "mourns the loss of nature itself. 'It makes me sick,' she says. 'The birds don't sing anymore. I no longer hear the song of the Chaco chachalaca bird. Even the jaguar that once scared me is suffering. That hurts me. I suffer from depression because of this. Now there is a hollow silence.

I feel as though our freedom has left us, has been taken from us with the nature that we have always protected.'"[18]

In 2013, a US National Research Council report, "Abrupt Impacts of Climate Change: Anticipating Surprises," called for massive investments in the investigation of pending instabilities that were not well understood. A new "abrupt change" climate science would identify signs thought to portend potentially catastrophic shifts in vulnerable Earth systems. While much progress has been made in predicting sea level rise, for example, the potential for destructive and surprising shifts, known as regime shifts, is now considered omnipresent. Regime shifts are "rapid modifications of ecosystem organization and dynamics, with prolonged consequences."[19] Some shifts are familiar: rising temperatures, stronger storms, increased fire frequencies, and decreased Arctic sea ice. Others, like the degree to which the loss of forest and ocean carbon sinks will affect temperature rise, are not known. Disasters linked to these shifts have been "more severe, longer, more frequent and less predictable than in the past."[20] According to the National Climatic Data Center of the US National Oceanic and Atmospheric Administration (NOAA), in the past decade, the United States has set new records for economic losses linked to droughts, wildfires, hurricanes, and tornadoes.[21]

In a period of runaway climate change, basic questions confront researchers—what combination of factors will turn a fire into a megafire? a storm into a superstorm?—which, in turn, raise questions about deploying emergency workers, for whom calculations of expected risks to life, based on current knowledge, are no longer accurate. The intensification of droughts, wildfires, hurricanes, and tornadoes does not follow linear schemes. Many shifts are not gradual but are happening abruptly or within shorter-than-projected time frames (within a person's lifetime, not centuries, and happening now). Jane Lubchenco, an environmental scientist, marine ecologist, and former head of NOAA, noted that these shifts were "harbinger[s] of things to come."[22] The models used to identify patterns and deviations may not be up to the task in this different world.

The message from scientists and policy makers has been that we need to expand our horizons to include an increase in consequential ecological shifts. Climate research has focused on producing linear extrapolations from recent history that imply a false sense of stability. The study of change in various fields, from biology and ecology to economics, is often predicated on the assumption that transitions are smooth. The statistical unlikelihood of stair-step, nonlinear, or abrupt transitions has lulled us into thinking they're impossible.[23] Gradual or smooth transitions are easier to project and thus can inflate a sense of control; for instance, in climate science, there has been an assumption that "slow [or gradual] processes pose small risks," and that "a choice can always be made to quickly reduce emissions and thereby reverse any harm within a few years or decades."[24] In economics, the notion of "incremental emissions" aligns with the notion that future damages can be projected and translated back to the present as an "appropriate damage value."[25] Such a value assumes that time can be bought, or that the level of destructive change is a choice (or that it is reversible). Guató Silva's sense of total loss has no place in a value system that says little about how Black, Brown, and Indigenous communities suffer disproportionate damage from the extractive industrial activities and environmental disasters that are a part of anthropogenic climate change.

Myths in the ecological sciences can also perpetuate notions of infinite adaptability. But, as evolutionary ecologist Simon

Levin told me, "there's no evolution working at [some] level to make sure that the [Earth] system is preserved. So, we may be filling up the atmosphere—with pollutants and greenhouse gases and things of that sort. Maybe we'll survive that, but maybe we won't, there's no [known] mechanism [to guarantee survival]." Amid apparently idiosyncratic ecosystem behaviors, and in the absence of such a mechanism, he suggested that "we may be adapting ourselves to extinction."

In other words, processes that are only partly quantifiable or entirely unobservable are driving a wider wedge between what we can observe and how we should act.[26] As damages grow, there will be less and less useful data to bring to bear on planning and preparedness, making anticipatory thinking that could guide transitions an even more difficult enterprise, as seen in the inaccurately predicted collapse of the West Antarctic Ice Sheet or in routinely miscalculated intervals of coastal flooding in the eastern United States. As the geoscientist James White put it, we will be left with knowledge gaps, or "areas of observation where we are largely blind," as well as more difficulty in determining where we are relative to dangerous thresholds.[27] In a public briefing on "Abrupt Impacts of Climate Change: Anticipating Surprises," the report mentioned earlier and that he coauthored, White illuminates the stakes of imagining paths forward as time runs out and climate futures start to rush in: "My hope is that we can study the planet well enough, monitor it well enough, understand it well enough, that we're not going to be blindsided. As a realist, I'm pretty sure we're going to be blindsided."[28]

Marine biologist Terry Hughes and colleagues saw this brink in warming tropical seas, where mass bleaching events in coral reef ecosystems have become too recurrent to give them time to recover. 1.5° or 2°C of warming (above the preindustrial level) means that every time such an event occurs, the bigger the "geographic footprint of recurrent bleaching" and so, the bigger the hole that is torn into "Earth's delicate web of life."[29] Worlds without coral reefs, the Arctic ice, and the Amazonian rainforest are no longer hypothetical. Ailton Krenak is a philosopher and Brazilian Indigenous leader of the Krenaki tribe, of which only 130 individuals are left. As he tells us, Indigenous people have confronted or "postponed" the disappearance of lifeworlds many times over.[30] Yet he redirects our attention to living futures that are still recoverable, not denied. In the words of the Brazilian Indigenous leader Davi Kopenawa—whose community has endured massacres, ongoing land theft, rancher-set wildfires, and violent resource extraction—"We all fear being crushed by the falling sky." Yet obligations to safeguard whole systems remain. Kopenawa implores his community and wider publics to "protect the forest. Prevent the river waters from flooding it and the rains from mercilessly drenching it. Repel the cloudy weather and darkness. Hold up the sky so it does not fall apart."[31]

These leaders' words reflect principles of safeguarding whole systems that have held despite ongoing violence and threats to self-determination. In this vein, Muscogee (Creek) geographer Laura Harjo writes of realizing futurity. Addressing seven-generation planning within Muscogee (Creek) communities, she defines futurity as "continually renovating our idea of the future and determining how to get there."[32] Indeed, futurity's effective schematizations are works in progress, a product of what she calls "wayfinding work." In what follows, I explore this work through the idea of horizoning, a mode of thinking and action meant to safeguard whole systems, even as the conditions for securing such outcomes shrink or disappear. I delve into how scientific experts and, in the second part of the book, non-Indigenous and Indigenous wildland firefighters contend with the pending horizons of abrupt change, with a specific look at efforts to contain wildfire, or, in the still-current metaphor, to fight against it. How do differently situated experts and eyewitnesses interpret possibilities for action, or, for that matter, what futility might mean? Holding back climate chaos, pushing back on extinction scenarios, holding up the sky—such images grasp the scope of the challenge: of protecting the circumstances of life so that they do not fall apart. Living with disappearance is not just a problem of more science, but of contending with dimensions of loss rarely seen.

3

When Paths Disappear

A sequence of images from a film called *Chasing Ice* shows one photographer's multiyear record of a bit of the planet shifting to a qualitatively different state. Atop thirteen glaciers, James Balog set up twenty-seven cameras that snapped photographs every half hour during daylight for several years. The stunning time-lapse images revealed, in Balog's words, "the horror and miracle" of the world's rapidly retreating glaciers.[1] They captured the dismantling of a stable Earth system occurring unexpectedly or faster than anticipated, otherwise known as an abrupt change. In the early 2000s, in eastern Antarctica, scientists "watched in amazement as almost the entire Larsen B Ice Shelf shattered and collapsed in just over one month."[2] The event was followed by the rapid and extensive Great Melt in Greenland in 2012. In 2020, two western Antarctic glaciers began breaking loose from their restraints; the loss of one of them, Thwaites Glacier, "could trigger the broader collapse of the West Antarctic ice sheet, which contains enough ice to eventually raise seas by about 10 feet."[3]

FIGURE 3.1. Retreating ice at Columbia Glacier, Alaska (stills from Jeff Orlowski's *Chasing Ice*, 2013).

Glaciers, along with ice caps, ice sheets, and sea ice, constitute a massive shield that reflects the sun's heat away from Earth's surface. Their essential cooling function is being compromised. A positive feedback, called the albedo effect, is amplifying the effects of climate change in a discrete chain of events. The more carbon dioxide is emitted into the atmosphere, the more heat is trapped on Earth. The more the ice melts, the more seawater, the dark color of which will absorb more heat. As this feedback becomes self-sustaining, the mirror-shield dissolves, seas rise, and the feedback incites other feedbacks in other Earth systems.

Cloud cover too acts like a giant shield. It reflects sunlight and, like ice, performs an essential cooling function. On a warming planet, the added sunlight coming to the earth's surface removes clouds, and cloud loss could add 8°C of warming within a century. A world without clouds begets a world without ice, which begets a world that is flooding and burning. Imagination is out of step with the dimension and pace of these feedbacks.[4] They are more like an indiscriminate wrecking ball upon Earth itself.

How are we to keep up with this continuous disruption of the Earth system? What metrical concepts apply to its chain of events? Before taking further stock of these changes, let me consider other more familiar depictions of how Earth systems stabilize or destabilize over time. Their variably sloping lines don't, however, tell the story of how the further behind schedule we fall in limiting emissions, the more the power of projection will fade.

The Fifth Assessment Report of the Intergovernmental Panel on Climate Change (IPCC) indicates four CO_2 concentration pathways through the end of this century (fig. 3.2). These pathways indicate a set of possible emissions trajectories for

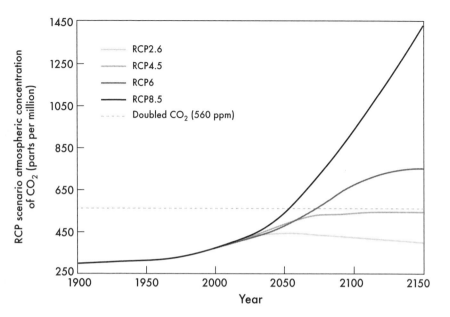

FIGURE 3.2. Representative concentration pathways of future CO_2 concentrations (lowest curve: best-case scenario; highest curve: worst-case scenario) (image drawn by Dana Nuccitelli, licensed under CC BY, http://skepticalscience.com/climate-best -to-worst-case-scenarios.html).

the purpose of assessing future warming scenarios. The lowest path represents a stabilized scenario; here, CO_2 concentrations would hover around 420 parts per million (ppm). That path is where the calamity of radical change in the climate system can be averted.

And staying on it has been the aim of intergovernmental bodies and campaigns. Until recently, policy making centered on insuring that the average surface temperature of the planet does not rise two degrees Celsius above temperatures in pre-industrial times (the IPCC uses 1850–1900 as the standard time reference). An IPCC 2018 Special Report announced the urgency of holding temperature rise at below 1.5 degrees

Celsius, a goal which requires that carbon dioxide emissions be cut in half by the year 2030. In recent years, atmospheric CO_2 concentrations have risen to 419 ppm, and those concentrations are rising. In 2020, global average temperatures reached 1.2 degrees Celsius above the preindustrial average, and "the monsters"—like firestorms, sea level rise, heat waves, floods, droughts, and rapid declines in biodiversity—have already come out.[5]

Each path is a statistical distillation of inputs and variables and a dim "analogue" of geophysical reality.[6] The planet's fate moves along paths with arbitrary endpoints and toward less or more risky scenarios. But these do not account for (carbon-cycle) feedbacks that exacerbate temperature rise from emissions. The scenarios hide magnitudes of peril that climate science cannot anticipate; they also expose the limits of models in delineating "how quickly very bad outcomes could show up."[7] Nonlinear temperature increases—which turn forests into kindling (so-called fuel) and create conditions for mountains in California and Colorado to burn when they should be covered in snow—form a trajectory in which what becomes of wildfire next is anyone's guess. On this altogether different path, wildfires intensify and pose complex operational challenges. They breach expectations of frequency, size, seasonality, or spread; notions of containment are upended by wildfires that grow out of proportion with—and can become largely independent of—current projections. As will be argued in later chapters, this other trajectory has yet to be empirically understood and effectively schematized. For now, let me turn to other visual representations that bring home the immensity of the stabilization challenge.

The concept of the "stabilization triangle," developed by Princeton University scientists Robert Socolow and Stephen

Pacala in 2004, was (and remains) an important heuristic for how atmospheric carbon dioxide emissions can be stabilized within a global management scheme (see fig. 3.3). The horizontal line of the triangle represents the most desirable emissions scenario, achievable within a fifty-year time frame ("the length of a career, the lifetime of a power plant, and an interval whose technology is close enough to envision").[8] In that time, we would ideally achieve the goal of "beating doubling," that is, of keeping atmospheric CO_2 below twice its preindustrial 280-ppm concentration. Assuming fossil fuels can be kept in the ground, aggressive mitigation will still be required to keep the line flat (and, ultimately, falling). To do so would mean cutting carbon immediately and with mitigation strategies that are available now. Letting up will make the target harder to achieve as emissions start to double or even triple (and so on) relative to preindustrial concentrations.

The lines transecting the interior of the triangle delineate "wedges of stabilization." Each wedge represents a specific strategy for reducing the rate of atmospheric carbon buildup. There is a wedge for increased vehicle fuel efficiency, wind and solar power, the reversal of deforestation, and carbon sequestration, for example. Maintaining a flat line, the path of stabilization, would be humanity's success story of creating a livable horizon for future generations. The longer the delay in enacting these strategies, the more humanity lives on borrowed time, the quicker the proverbial sky will fall, and the more extreme, risky, or futile the measures needed to achieve stabilization will become.[9]

Stabilization, in fact, is a moving target—an outcome of mitigation opportunities that have either been taken or squandered, and an accounting of how much of the carbon problem is being passed on to future generations. In 2011, when one of

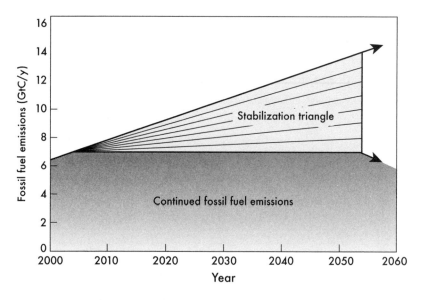

FIGURE 3.3. Stabilization wedges, 2004 (Pacala and Socolow 2004).

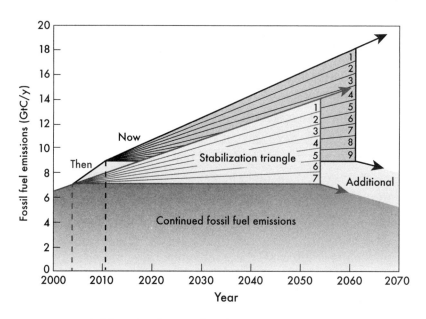

FIGURE 3.4. Stabilization wedges, 2011 (Socolow 2011).

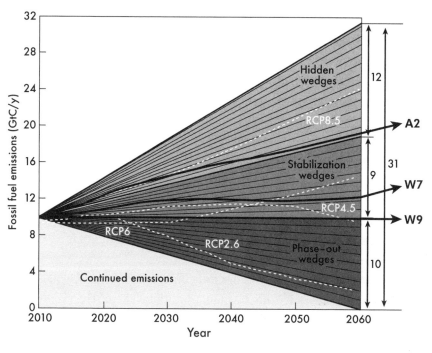

FIGURE 3.5. Stabilization wedges, accumulating cost of delay, 2013 (Davis et al. 2013).

the authors (Socolow) revisited the stabilization triangle, he added more wedges to achieve the same target and in order to compensate for the additional unchecked greenhouse gas emissions in the intervening years (see fig. 3.4). Inertia, intransigence, and the politics of old energy technology got in the way, so much so that "nine wedges [we]re required to fill the stabilization triangle, instead of seven." Two years later, a 2013 publication suggested that eliminating emissions over fifty years would require no fewer than nineteen wedges: nine to stabilize emissions and ten more to completely phase them out. If decarbonization does proceed quickly enough, the authors wrote, twelve "hidden" wedges will also be necessary, bringing the total number of wedges to thirty-one (see fig. 3.5).[10]

The fifty-year horizon, employed as a pragmatic tool to reorient the policies and investments of modern energy consumption, hit a crucial limit as the stabilization triangle started to tip over itself. The ramp of the doubled, tripled, or quadrupled atmospheric CO_2 emissions concretizes a costly delay in which the idea of living on borrowed time is no mere metaphor. The wedge becomes a parable of sorts, about a further unmooring from safety: a situation, not of returning to past CO_2 levels, but of never catching up to an ever-receding horizon of possible stabilization.

In this situation, damage increases as the ability to reckon with its scope decreases. We begin to face assemblages of accumulating surprise, consisting of overlaid events whose physical interactions and net impacts are unknown. In characterizing key scientific challenges in the wake of Hurricane Sandy, among the costliest hurricanes in US history, Marcia McNutt, the former director of the US Geological Survey, stated: "It is not the gradual rise of sea levels that is going to get anyone. It is the combination of extreme events superimposed on that gradual rise . . . that will destroy our natural protection and offer much less protection for future storms."[11]

In other words, any estimation of destruction in the future must be recalibrated in accordance with shrinking baselines of protection for entire ecosystems. "We have already crossed a threshold. Superstorm Sandy was a threshold, and we crossed it," McNutt noted.[12] This threshold also represents a moment when human projective powers falter. What experts are used to seeing as a problem of uncertainty that can be resolved with more data may, in fact, be a complex system on the verge of collapse. This sets the stage for surprises that our present tools are simply unable to manage, let alone suppress.

In McNutt's rendering, the climate change "monster" isn't so much a singular phenomenon or event as it is an ongoing process of destabilization in which scientific observers can no longer establish a reasonable baseline for projecting future patterns of change. As one engineer working with coastal scientists to improve infrastructures on the US North Atlantic coast suggested with regards to preparedness for floods and torrential rains, "there is a switch beyond which we cannot plan" for these events. As climate-linked changes outstrip the capacities of our infrastructures and become the norm, they rob us of critical time: "It went so fast," said a resident of a German village that was destroyed by floods that overwhelmed the country's flood alert system in 2021. "You tried to do something, and it was already too late." His words speak to the importance of enhancing infrastructure for "previously unimaginable volumes of rain," but also to the limits of such enhancement in environments that increasingly deny both a foothold and the chance for an effective projection.[13] In what follows, I ask how models of knowledge have tried to make sense of this pattern of denial through the work of "regime shift" science.

Earth as Tipping Place

In 1983, the lake ecologist John Magnuson and colleagues coined the phrase "invisible present" to flag problems in scientific observation of change in natural worlds.[14] He had been serving as the director of the University of Wisconsin–Madison's Center for Limnology, whose laboratory sits on the shores of Lake Mendota. This is the same lake that Stephen Forbes, the aquatic ecosystem science founder, had once called a microcosm. In it, fish were "remarkably isolated" and unable to sense impending doom. For Magnuson, it was not the fish but the scientists who were stuck in a microcosm in which some processes of change are "hidden from view or understanding." That is because the scientific modus operandi can itself either hide them or overdetermine what is observed. An analysis of lake sediments representing "hundreds to thousands of years" will yield only "a coarse history of past climatological events." Natural worlds that seem static when observed over a decade are, in fact, on the move when observed over a century.[15]

The choice of temporal frame becomes critical to aligning expectation with what we see: "certain biological and physical processes simply take time"; there are lags between the chain of events that inform these processes (say, the duration of ice cover on a lake or a change of seasons of plant flowering) and our observation of them.[16] Such lags make it possible to pick up some patterns, but not others. Phenomena that might have been occurring gradually and then shift abruptly constitute change that is literally without dimension.

The invisible present becomes a kind of scientific nowhere, in which observations of change can be out of sync with physical reality ("even when observed, many of these changes are understood by no one").[17] Runaway change, in which baselines constantly shift, represents its most extreme unmoored state. It takes an "unusual person" to do science here, in conditions that are shot through with potential for misinterpretation.[18] Such potential will be a critical theme in upcoming chapters. Indigenous fire managers will talk about fire management that created landscapes that are now incompatible with fire, the result being an onrush of uncontrollable fires that are destroying valued cultural sites. Not only are baselines changing, but as Jack Cohen, a retired research physical scientist with the Rocky Mountain Research Station's Fire, Fuel and Smoke program, put it to me, "we will need more knowledge of how not to break things even further." If certain actions taken today "break" rather than preserve options for sustaining life tomorrow, then, in his words, "you are going to see something disappear." Responding to this crushing invisible present cannot be left to one set of experts, but must be undertaken by an array of interdependent knowledge holders who, in their partial comprehension of patterns and shifts and how they occur, can enlarge the breadth of resources with which options are preserved rather than broken.

Early in my attempt to understand environmental instability and its links to anticipatory thinking in science, I reached out to American lake ecologist Stephen Carpenter, who joined Magnuson at the University of Wisconsin–Madison's Center for Limnology in 1989. I wanted to talk with him because of his groundbreaking work on ecosystem dynamics, lake experiments, and environmental management (for which he received the Stockholm Water Prize in 2011). Carpenter has collaborated with an interdisciplinary group of scientists, well known in their respective fields (which include shallow- and deep-lake ecology, coral reef ecology, and Earth systems, among others), in quantifying critical thresholds, or tipping points, for different ecosystems under threat. He and his colleagues have pointed to the limits of analytical approaches to complex ecosystems; these can create what they call "spurious certainty."[19]

By filtering out poorly understood variables, spurious certainty can engender erroneous images of the natural world at the cost of a "better estimate [of] our uncertainties." Too often, politicians have relied on such certainty to project control over a situation when it is out of control, or to limit public debate about or hide the costs of increasing complexity. With spurious certainty, short-term horizons are the rule. As we will see, in the United States, such a trend has favored aggressive wildfire control in which various interests (political and financial) frame explosive wildfires as merely "episodic" or aberrant in an otherwise stable trend line.

Such trend lines suggest constant rates of change. In a form of spurious certainty, their linearity can sometimes smooth over shocks and perturbations (linked to various forms of human exploitation, including releases of anthropogenic greenhouse gases), producing a mirage of stability. With every smoothing over, course correction comes at a steeper price. As has been witnessed in record-breaking wildfire seasons in the western United States and in the world, the reality of destabilization can no longer be subsumed in a mirage. Considerations of how ecosystems break under anthropogenic pressures have to be coupled with what models do to the process of observing, and what they can hide.

Carpenter and colleagues have been especially concerned with modeling how large anthropogenic shocks can change an entire ecosystem. He was a graduate student in the 1970s, when the equilibrium models that had dominated ecology for decades were unable to account for key ecosystem dynamics. As he told me in a conversation, in ecology, "everything that is really interesting" occurs in relation to transitions, linked to external shocks, internal instabilities related to feedbacks, or the entry of new actors (like invasive species) into a given ecosystem. "These [transitions] really are the important events, and the dynamics around equilibria really are not."

Many ecosystems can and do undergo rapid modifications in their structure and function. Even though such regime shifts are common in nature, environmental managers still have a hard time admitting that their outcomes, no matter how unusual, can't be managed with usual models and tools. In Carpenter's view, "people differ enormously in their tolerance of uncertainty. To think broadly about the future, you've got to have considerable tolerance of uncertainty, because if you're not tolerant of it, you end up just trying to canalize your planning into some very narrow groove." The narrower the groove, the narrower the chance of adaptation, and the more we find ourselves in "areas of observation where we are largely blind," in James White's words. Carpenter wants tolerance for uncertainty to become "automatic and common" and a source of creative problem-solving, All too often, however, when environmental management agencies want to believe "that something is going to work" and that what they're doing is an answer and not a question, or a hypothesis, they reiterate spurious certainty. Their solutions will end up becoming traps, and public expectations "will almost certainly be dashed."

One-Way Trips

The idea of nonequilibrium in ecosystem dynamics has several origins, among them the work of C. S. (Buzz) Holling, who elaborated on the "stability of ecological systems" and communicated the concept in simple ways. It was also explored by a less frequently cited theorist, the mid-twentieth-century French mathematician René Thom.[20] Thom won the 1958 Fields Medal for his work on the arcane subject of topology: the mathematical study of how objects preserve their formal properties when deformed (the word is from the Greek *topos* or place and *logos* or study). A ball can be squashed into a bowl; a sphere can be stretched into a cube. Because of what Thom called "structural stability," objects can retain certain topological properties. They can also carry the shape of things to come.

Thom was interested in what allows these properties to carry over, as well as behaviors that occur when they don't.

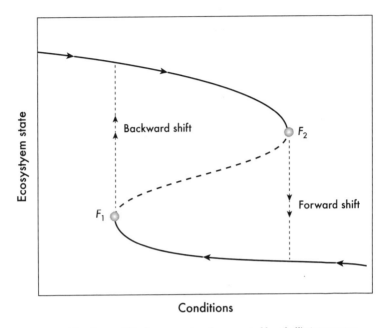

Conditions

FIGURE 3.6. The abrupt shift of an ecosystem (represented by a ball) tips across a threshold (F2) and into an alternative stable state (F1) (Scheffer et al. 2001:592).

The question of thresholds and breakaway states is at the core of regime shift science (see figs. 3.6 and 3.7). Imagine a ball sitting on the bottom of a basin that can be said to be in a state of rest (a stability state). An outside shock, if strong enough, can push the ball across a tipping point (F2) toward an "alternative stable state," whose topological properties cannot be known in advance.

Let's now think of the ball as an ecosystem. Depending on the amount of pressure (say, from the level of greenhouse gas emissions), the ball steadily moves along one curve. At F2, it "freefalls" along the dotted line and into another stable state (F1). Once that fall occurs, a path back will be unavailable, "no matter what actions human societies might take."[21] The message

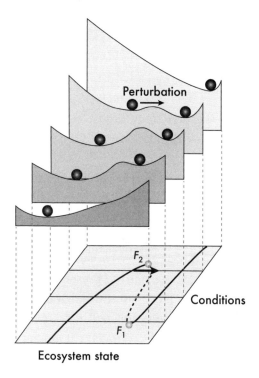

Perturbation

F_2

Conditions

F_1

Ecosystem state

FIGURE 3.7. Representation of threshold (F2) demarcating an ecosystem's shift to an "alternative stable state" (F1) (Scheffer et al. 2001:592).

conveyed by the simple graphs is that the longer cutting emissions is delayed, the more imminent the instability, the narrower the options for averting it or responding to it, and the more imminent an alternative stable state will be.

For Carpenter, Lake Mendota has been a place to test the theory of "alternate stable or quasi stable configurations" against field observations. He and his colleagues have discovered where critical thresholds in the structure and function of lake trophic systems ("food webs") reside, and how, he said, "with enough pressure, you can move an ecosystem across a threshold into a different configuration."[22] A tipped lake is typically what is known as a eutrophied lake, a stubborn environmental problem that results from fertilizer runoff (nitrogen and phosphorous) that feeds excessive plant growth (and decomposition) in a lake, producing too much CO_2 and robbing the lake of oxygen. Lakes become smothered in blankets of thick algae, which means that food production and "the whole metabolism of the ecosystem changes."[23]

In theory, the change is reversible, says Carpenter. But here's the rub: once the lake is over a threshold, "eutrophication is a one-way trip."[24] The alternative stable state—in this case and, potentially, in other ecosystems—would be permanent. Before these thresholds are reached, there is much work to do in recovering time to act and preserving the conditions of life. This means building institutions that can learn, evolve, and maybe somehow not only forestall water quality problems, but also hold onto more or less familiar wildfire regimes. Later, I'll present the work of scientists and emergency responders who are planning for broader regime shifts without necessarily knowing how to predict wildfire severity from one year to the next. Now I want to consider such immediate data-scarce environments in their own right: not as knowledge voids but as experiments in capacity building, of horizoning, in which a larger reckoning with runaway climate change can take place.

4

Horizon Work

Across time, people have used horizons as points of reference in navigating physically incoherent worlds (seas, for instance). The word "horizon" derives from the ancient Greek ὁρίζω (*horizō*), meaning "I mark out a boundary," and from ὅρος (*oros*), meaning "boundary" or "landmark." Renaissance architects contrived horizon lines to properly orient objects in three-dimensional space. Early modern surveyors devised mercury-filled "artificial horizons" to create images of a level surface, against which the "inconstancy of the terrestrial horizon" could be judged.[1] Today, robotics engineers encode "predictive horizons" in remote machines (such as extraterrestrial rovers) to allow for autonomous self-correction in the navigation of craterous conditions on Mars. In meeting such course-plotting challenges, data from the past is useful, but only up to a point. And precisely when data is no longer useful, and prediction capability derived from past or present information becomes misleading (or yields high computational cost or instability), a new predictive horizon must be put into place.[2]

In these examples, horizoning is a distinct kind of intellectual and practical labor undertaken in conditions in which the fate of entire "systems" (a boat or a rover) is at stake.[3] In such environments, it generates projections whose accuracy is tested in the movement they afford. It is also a practice of continuous self-correction vis-à-vis changing baselines of safety and knowable risk. It can yield scaling rules for identifying and "maintaining a safe distance from dangerous thresholds."[4] In conditions of extreme ecological uncertainty, horizoning entails a fine-tuned awareness of a system's exposure to jeopardy, without which operators will be flying blind.

As these processes suggest, horizons are not open-ended or metaphysical. Rather, in circumstances with limited visibility, they call out differences between meaningful and baseless projection. They acquire value according to their ability to retool complexity, making it navigable. As an anthropologist once wrote of the Poluwat atoll navigators of Micronesia, "[T]he sea is a demanding master. No style of thinking will survive which cannot produce a usable product when survival is at stake."[5] Seafarers have applied related expertise, including the use of sensory parameters—the smell of a certain forest, certain patterns in tidal waters—to produce "usable" horizons. Trapped in dense fog or storms, they deduced their positioning vis-à-vis a previously estimated location or fix. Their so-called dead reckoning "makes good" on faulty or fleeting information, thus allowing movement forward or preventing a crash (or disappearance) of an entire system.[6] The fact that entire trajectories, systems, or worlds may be at stake is precisely what makes horizons so real.

In what follows, I continue exploring efforts to apprehend runaway climate change as a complex phenomenon. Where some see sudden or dramatic processes that are irreversible, others see spatial and temporal dynamics and abrupt transitions that have been poorly horizoned so far. To illustrate the latter, I consider small-scale experiments, where the identification of so-called tipping points raises key questions about action-under-uncertainty in broader ecological fields. Such experiments lay out the challenge of practical intervention, and of how stabilization as a shared goal can be reckoned with.

Crossing Tipping Points

Marten Scheffer, a Dutch lake ecologist, picks up and jiggles a flask of daphnia, one of hundreds filled with zooplankton and cyanobacteria arrayed on white shelves in his laboratory at Wageningen University in the Netherlands. We have established a great rapport over the last two hours. "Daphnia," he tells me. "Water fleas?" I respond, before we walk toward a piece of laboratory equipment. It is called a microcosm: in this instance it is a small, steel-enclosed, climate- and light-controlled model of a lake ecosystem. "It's the ICU of water fleas," he says. With its attached laser beam and algae-fluid container, the "ICU" is said to contain a complete biophysical system whose dynamics, he noted, "we like to think of as mimicking the dynamics of the Arctic ice sheet." I wasn't prepared for this scale shift from metal container in a Dutch university town to a polar ice sheet in the northernmost region of Earth. I soon learned that something hard to see is being modeled in the microcosm: an invisible present of confounding feedbacks and the abrupt shifts that they can induce.

Scheffer is world renowned for his studies of tipping points, regime shifts, and alternative stable states. Early in his career, he showed how such threshold-crossing "critical transitions"[7] occur in shallow lakes. Based on that knowledge, he has turned turbid lakes into clear ones using novel strategies. He has been a leader among others in the experimental testing of early warning indicators that can help anticipate when ecosystems are approaching such transitions.

The past and present are full of evidence of abrupt and irreversible shifts occurring without warning or signs of proximate cause. The Sahara Desert was made up of numerous wetlands six thousand years ago, before it suddenly became a desert. The Caribbean coral reef recovered from excessive fertilizer runoff until it did so no more. (Algae overgrowth linked to fertilizer runoff choked off needed light for coral larvae settlement, disrupting a crucial phase of coral life history.) Other current examples include the potential shutdown of the Atlantic thermohaline circulation (the ocean's "conveyor belt"), the "dieback" of the Amazon rainforest and boreal forests, and the loss of the Greenland ice sheet.

Back at the microcosm, the stage has been set for the inquiry into feedback systems. An initial experimental setup puts two organisms (pairs of algae species, in this case) in competition for the same resources, like light or nutrients. In the resulting stable state, the rules of mutual competition are identified. This baseline provides the starting conditions for manipulating rules and observing shifts toward different community structures, volumes, or densities. For example, increase the temperature and the algal fluid becomes darker, a sign that the structure is changing. If light is increased, one of the pair of algae species will begin displacing its competitor for resources.

Experiments in microcosms allude to tipping point dynamics in polar ecosystems, where algae grow. Their dark color attracts solar radiation. Thriving on photosynthesis, they will become ecologically dominant. As competition ramps up for the same resources, they will start to displace their ecological competitors (such as invertebrates living under and above the ice). Competition produces a positive feedback: more solar irradiance means more competition that, in turn, means more algal dominance and hence accelerated ice sheet melting. A whole system moves toward a tipping point, reaches a brink, and cascades toward a different set of ecological rules.

To be sure, smaller ecological and biotic successions create this bigger event. Yet their unique temporal and spatial unfoldings cannot be reduced to that event or to a general "point," like a tipping point. Nor are their end-formations in lockstep with their causes, which can themselves result from constellations of interacting causes or other unknown feedbacks.[8] Thus abrupt ecosystemic shifts point to a present that becomes an invisible present. To borrow a term from the exploration rover, they have limited predictive horizons.

But events inside the microcosm are said to offer clues for making sense of these shifts. For instance, as the model algal communities are perturbed under controlled conditions, they will take longer to bounce back to a resilient state, a telltale sign of an impending regime shift. They will start "flickering" (an actual term), oscillating between alternative stable states,[9] before reaching a phase of "critically slowing down." Once past a certain threshold, a transition becomes a cascade into a new stable state. What looks from the outside like a surprising shift (here an algal community "jumps" into a new state) can be anticipated by patterns of flux on the inside.

At larger scales, abrupt shifts are becoming routine. In the Arctic, the overgrowth of sunlight-dependent marine algae is devastating polar food chains. Move over to the ocean: once-protected coral reef ecosystems are dying as ocean waters acidify from carbon dioxide oversaturation. Now over to land: after certain temperature thresholds have been surpassed, the production of staple foods such as corn in some parts of the world is at risk. The work of tipping points can be both alarming and mundane in the sense that tipping points can lurk in everyday objects and infrastructures. Automotive brake, bridge, and energy distribution systems are built to accommodate specific "loads" of speed, weight, wind, or temperature; these can accumulate and "destabilize" a system past some critical threshold. While engineers have designed anti-lock brakes, preventing cars from spinning out of control, the same cannot be said for ecosystems; the science underpinning nature's abrupt shifts has a way to go in preventing the unwanted collision.

Catastrophism

While the conceptual chasm persists, the idea of abrupt shifts in nature is not new. Speculation about them stretches back in time and reflects a group of related theories, such as the eighteenth-century ideas of "saltation," which posited sudden and dramatic change (from the Latin *saltus*, or jump), and "catastrophism," which dominated the study of life's history in the early nineteenth century. The French naturalist Georges Cuvier saw catastrophism in fossils, arguing, for example, that fossil elephants "prove[d] the existence of a world previous to ours, destroyed by some kind of catastrophe."[10] By the early twentieth century, ideas about sudden transitions and inexplicable jumps were downplayed in the Western scientific canon. Charles Darwin, inspired by the gradualist uniformitarianism of geologists James Hutton and Charles Lyell, declared that—just as it abhors a vacuum—nature abhors a jump.[11] More heterodox theorists were not happy with the reduction of the study of complex life-forms to a monotony of "filling in the blanks" in stories of minute and gradual variation.

Among those keeping the possibility of abrupt shifts alive was Scottish biologist D'Arcy Thompson, who rejected this vision, writing that to "seek for stepping-stones across the gaps between is to seek in vain, forever."[12] Best known for his studies of morphology (the study of the forms of living organisms and how they change), Thompson saw organic forms as derived from the imprint of physical forces. His 1917 book *On Growth and Form* is filled with examples of this imprint. The perfect hexagons of honeycombs, the supposed paragons of bees' naturally selected instincts, were, in his view, a physical result of uniform compression on circular cells. A femur bone's underlying array of supporting bony trabeculae reflects the stresses of physical forces and could not be a result of inheritance. In such "direct molding," there is no hidden purpose (or teleology) in how a natural form should evolve, only the "plainest principles of mechanical causation" in how it *can* evolve.[13] Formal properties (and how they carry over, or don't, from one species

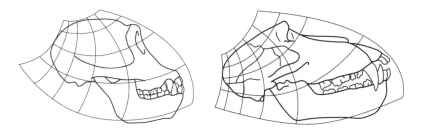

FIGURE 4.1. Transformation of a chimpanzee to a baboon skull (from Richards 1955:458, modified from Thompson 1942).

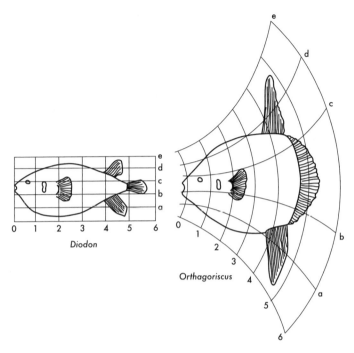

FIGURE 4.2. Transformation of the pufferfish *Diodon* to the sunfish or "mola mola" *Orthagoriscus* (from Richards 1955:460, modified from Thompson 1942).

to the next) can be anticipated, such that a puffer fish will be a puffer fish and not an ocean sunfish; a chimpanzee will be a chimpanzee and not a baboon, and so forth.

This speculative thinking about morphological shifts never evolved into a research program. With the rise of biological (and later, molecular-genetic) techniques, morphogenesis itself became somewhat of an intellectual backwater as evolution, with population genetics at its theoretical center, became a study of minute interchangeable parts (changing distributions and frequencies of alleles, etc.). In this new context, *On Growth and Form* was deemed an "unusable masterpiece."[14]

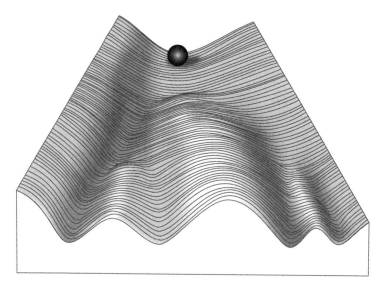

FIGURE 4.3. Epigenetic landscape (Waddington 1957).

But it did inspire other significant twentieth-century inquiries into the conceptual chasm of abrupt change, including those of British developmental biologist C. H. Waddington. Working to reconcile an organism's genes and its form, Waddington founded the field of epigenetics (or "the branch of biology which studies the causal interactions between genes and their products, which bring the phenotype into being"), developing a string of concepts that are critical to biology today.[15]

One of those concepts is the epigenetic landscape, which visualizes how genes and environments interact in developmental pathways. Let's imagine that the ball at the top of the landscape (see fig. 4.3) is a cell or a seed. As it moves down one valley and not another, its developmental fate unfolds. The terminus of each valley represents a certain outcome of differentiation. A cell becomes a specialized organ; a seed, a particular tree species. If conditions change (expressed in any slope in the landscape), the cell that was headed toward becoming one cell type undergoes an abrupt shift and becomes another type.

With its valleys and ridgelines, the epigenetic landscape echoes the curves of regime shifts (see fig. 3.6). Its lines represent a topological universe of stabilities and instabilities, with rules of biological assembly that a more conventional evolutionary logic cannot accommodate.

This other evolution fascinated René Thom, the mathematical topologist whose work I elaborated on earlier. He introduced a conceptual (if somewhat florid) nomenclature for its dynamic swings and fluctuations. Where a cell becomes a heart cell, and not something else, he called it a "catastrophic bifurcation." Bifurcations were waiting to happen everywhere, even in what he called the "most homely" of phenomena.[16] His mathematical modeling of surprise and the shape of things to come in indeterminately evolving systems came to be known in the 1970s as "catastrophe theory." The theory is now used by many scientists to mathematically describe something as benign as how, let's say, this type of snail shell is not that type of clam shell, or something as terrifying as how, without much warning, familiar habitats and forms of life can give way to entirely new or unfamiliar ecological states.

Regime shifting, catastrophic bifurcations, and alternative stable states signal rearrangements in ecosystem behaviors that defy conventional expectations.[17] But even these ideas, along with more popular variations on them, like tipping points, fall short. A messy, frayed vastness of causally intertwined possibilities that make up a broad, undulating border of tipping points is less like a ridge than a cliff. Depending on various factors, such as wind conditions, some possibilities will be realized sooner rather than later. Others might require

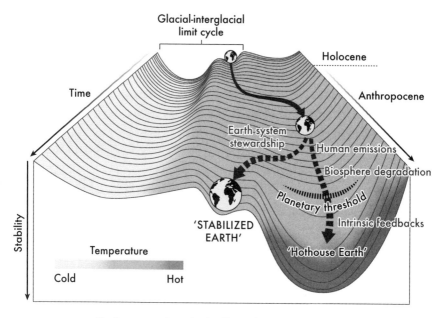

FIGURE 4.4. Earth system trajectories (Steffen et al. 2018).

circumstances that take much longer to develop. Thus this edge contains a huge range of events-in-waiting, all subject to differently evolving and disintegrating stability states.[18] We stand in a space between the present and the future that is liminal and porous, and in which critical ecosystems are disappearing or under threat. Put differently, in the gap between tipping points and "new states" is a series of un-happenings: something isn't happening that should have been happening, and cycles of bounce-back stop happening or are choked off by other kinds of cycles or forces (in an active diminishment of resilience). These accumulating interactions, the subject of an influential essay on planetary thresholds, separate a stable Earth from a "Hothouse Earth" with "new, hotter climatic conditions and a profoundly different biosphere." Waddington's epigenetic landscape inspires a model of these interactions.[19]

Today, abrupt ecosystem shifts present a problem of projection into the shape of evolutions to come. CO_2 concentrations, rising temperatures, and biotic shifts do not share a single threshold, and many ecosystems do not or may never have demonstrable tipping points. Even so, many scientists argue that we would be better off if they did because knowledge of tipping points underscores the cost of damage from inaction. "[There] is no point in discovering the precise tipping point by tipping it," as two prominent climate scientists write, referring to an Amazonian tipping point and the need to stop deforestation.[20] Twenty percent of the Brazilian Amazon has been deforested since 1970. Logging and rancher-set forest fires burning out of control in changing climate conditions are threatening another 20 percent. At 40 percent, the "forest will be lost forever and replaced by savannahs" as precipitation decreases.[21]

Importantly, tipping point research in tropical forests, oceans, and lakes diverges from an era of science in which assessments of climate chaos were overly conservative (and "larded with caveats"), or new disasters were regarded as the latest wake-up calls.[22] Although the concept of tipping points is fraught with unreliable estimations, maybe our mental models of end-states—let's call it extinction—are the real problem. Maybe the edges of extinction are somewhat malleable if only the right models and intuitions can be applied.

Flipping Ecosystems Back (a Cautionary Tale)

A question lingers: If ecosystems move past tipping points, can one imagine flipping them back? To consider this question, I return one more time to small-scale lake experiments, where theories, when tested against field observations, sometimes point to workable schemes. Shallow Dutch inland lakes were once inundated with nitrogen and phosphates from local agricultural fertilizer runoff. Clear lakes had turned turbid—the inevitable fate of all lakes that are termini of fertilizer runoff. If shallow lakes could be tipped back, what would this portend for restoration efforts in other ecosystems, such as troubled oceans or tropical forests?

Scheffer worked as an aquatic manager early in his career, when predominant equilibrium frameworks were losing theoretical traction. Referring to René Thom's theory of alternative stable states, he said, "There was a theory around, but no one could really show how it worked. The theory resonated with an intuition that people had, but no one could show it in practice." Experts working to restore aquatic ecosystems were often misguided by certain expectations. A fellow ecologist thought that once an aquatic system was restored, it would stay that way, and that "basically it could not be destabilized." That ecologist had carried out experiments in canals and ditches to prove his belief—for example, he "put lots of fertilizer into those ditches, and they never became turbid." This ecologist "would have been right for ditches but not for lakes."

While ditches or canals ("small systems") have many lake-like properties, there are some essential properties that they do not have, like wind and wave effects. Lake dynamics are also influenced by predatory fish that stir up sediment when they move. Muddy waters stop sunlight from reaching lake-bottom plants. These plants can't grow, nor can they provide hiding places for zooplankton (small drifting organisms that filter water). Zooplankton become exposed to the predation of fish, so the water will not be filtered—it will stay turbid, no matter how much fertilizer is removed.

Without an understanding of these interacting phenomena, quick-fix attempts at reversing a lake's turbidity would certainly fail. Scheffer opted to restructure the lake's ecosystem (a process called biomanipulation). The thousands of predatory fish that kept the lake in a mucky state were "experimentally removed" through seining and trawling. This halted the water turbidity, allowing light and oxygen to help reestablish aquatic life. Of course, one might raise ethical questions about nonlake inhabitants making decisions on behalf of those in the lake. But in flipping terms, Scheffer told me, the results were "quite spectacular . . . it was relatively cheap and easy to do, and there was a theory behind it. . . . And so, then, you have all the elements for [the theory's] rapid spread."

This approach to pushing shallow lakes out of algae-covered and oxygen-deprived states has become standard. It seems hopeful enough. But it is more a cautionary tale about the pitfalls of stabilization exercises at larger scales. A lake is a relatively circumscribed entity affording some experimental control. The same cannot be said for larger aquatic systems or for Earth's climate system. Yet the imperative of mitigating the worst outcomes of global warming is increasingly pointing to geoengineering, a planetary-scale intervention in offsetting warming. It involves planes injecting billions of tiny, reflective sulfur dioxide particles into the upper atmosphere, producing chemical clouds reflecting sunlight. But there isn't, and perhaps will never be, enough control to deal with the potential adverse outcomes of this experiment. Moreover, while solar geoengineering might temporarily reduce global average temperatures, it is no remedy for irreversible climate destabilizations that have already been set in motion. The prospect of this hack raises a critical question: On whose behalf are large-scale experiments being made, and who will pay the price of the spurious certainty that underpins them?[23]

Since my conversations with Scheffer, boreal forests just south of the Arctic Circle have gone up in flames. The 2016 Fort McMurray wildfire in Alberta, Canada, forced eighty-eight thousand people from their homes. Burning over a two-month period, it released the equivalent of 5 percent of Canada's annual greenhouse gas emissions (forty-one million tons). In 2017, British Columbia wildfires tripled that province's annual carbon footprint. The 2019 Arctic wildfires released more CO_2 in one month (fifty megatons in June) than Sweden did in that entire year.[24] Later in 2019, apocalyptic Australian bushfires emitted four hundred megatons of carbon dioxide into the atmosphere.[25] Brazil, meanwhile, is the scene of a vast environmental crime. A vital carbon store is being destroyed by rapid deforestation, logging, and a massive number of illegal fires.

In this century's first decade, a coalition of political, scientific, and economic actors pulled the Brazilian Amazonian rainforest back from the brink. They did so by imposing critical moratoria on the soybean and beef industries. They used satellite networks to spot instances of illegal clearing and burning. Scientific agencies worked with judiciary systems to strengthen environmental enforcement and commodity supply chain traceability, holding perpetrators and sellers of deforested land products accountable.[26] They also strengthened and helped demarcate territories where Indigenous peoples and other communities, threatened by logging, mining, and ranching, are the frontlines of defending rainforests. It wasn't geoengineering but coordinated acts of stabilization, based in political will, human rights, and technical expertise, that allowed Brazil "to reduce forest clearing in the Amazon by an astounding 70 to 80 percent."[27] This decade-long effort made the forest available to do its critical work of absorbing the key driver of global warming, carbon dioxide. That know-how remains available and ready to be deployed again, in Brazil and elsewhere.

Today, wildfires are a climate emergency. In what follows, I dive into this crisis as it comes to a head in the south and northwestern United States. Given the deleterious forces that have so radically transformed the planet that humans now have their own geological epoch, the Anthropocene, the question is not how wildfires should behave, but how, given those forces, they can behave.[28] I ask wildland firefighters, managers, and fire scientists about the meanings of stabilization and how, in the unhinged ecological fields in which wildfires now burn, stabilization as a shared goal is either thwarted or achieved.

5

"Throw Away Your Mental Slides"

In what is now a standard anecdote about expert intuition in extreme circumstances, a firefighting commander leads a crew into a burning house. The crew points a water hose at a fire in the back of the house. The commander can reasonably expect that the temperature inside will drop. Instead, it gets hotter, so he orders his crew members to evacuate. Within seconds, the floor they had been standing on collapses (from a fire that, unknown to the crew, was burning in the basement). The commander was able to "think fast" (or make that split-second evacuation call) because his experience with fire patterns allowed him to sense the anomaly.[1] Put differently, he could keep the gap between what he thought was happening and what did happen as small as possible.

Now imagine that the basement fire is hotter and spreading faster than any fire the crew has experienced before. What if a different leader doesn't have the means or experience to act on a similar intuition—to make the right split-second call to protect the crew? In the case of a wildfire, what if the crew leader has as much experience as anyone could possibly have, but it still isn't enough to enable intuitive fast thinking? In any significant wildfire incident, there will be plenty of things that the crew can't see, anticipate, or intuit: the random gust of wind carrying a burning ember across a highway and into a suburb built next to undeveloped wildland, sparking a new fire, or into a nearby mountain range, igniting drought-parched vegetation in one of its canyons.

Now, imagine one of her crew members, who happens to be in a neighboring canyon, assessing structures and properties that may be at risk, sizing up a fire, or looking for potential locations for digging a barrier or fireline. The crew leader is expected to send people on scouting missions like this one. The canyon has turned into a wind tunnel. A creeping little fire is just one gust away from becoming a fireball. The crew member will not see the fireball coming; he might hear a sudden roar and not know where the sound is coming from. As he struggles to gain his footing, the fire overwhelms him. No leader or crew should be put in a circumstance that no longer correlates with the fires they've learned to fight. Yet, every day, firefighting crews are expected not only to face rapidly changing fire behaviors, but also to rewrite the rules of their own expertise while doing so. Meanwhile, the public remains mostly oblivious to the nature of this heightened risk.

We want the assurances of the world of the first commander, in which the consequences of disaster can be contained and expert intuition works.[2] A mental model of fire enables a sensing of patterns and deviations. In this world, the public delegates emergency response to a group of seasoned professionals, and tomorrow, for the most part, will look like today. But we now live in the world of the second commander, where conditions no longer lend themselves to such expertise. The mental model does not work. The gulf between what is predicted and what actually occurs has grown too large, even for the experts.

The pages that follow probe the consequences of uncertainty in this emerging other world, in which mental models are undermined by new circumstances. In this world, better models of the physical realities that surround us are needed. Firefighters told me there used to be enough constancy in wildfire's shapes and behaviors to afford some pattern recognition. They carried a deck of stored images, what they call mental slides, of previous fires. Like the first commander, they instinctively called upon these slides to discern the behaviors of new fires. Today, relying on those slides can get in the way of that discernment. Their trainers urge them to throw away their old mental slides because trust in patterns has become an occupational hazard. As they throw away their mental slides, what are they supposed to replace them with?

By the time it had ended, 2015 was seen as the worst wildfire year on record in the United States. Wildfires consumed ten million acres and fighting them cost taxpayers $2.6 billion. In California alone, there were about thirty-four hundred wildfires, a thousand more than the average over the previous five years. Fire seasons are becoming much longer (on average, over one hundred days longer than they had been five decades earlier). For two weeks in 2015, recruiters in the national dispatch and coordination system hit a resource limit: there was simply nobody left to recruit to fight the fires. Military personnel, volunteers, and even prisoners were conscripted into emergency response efforts. The truth is that it is hard to find stability in numbers. In 2017, dispatch services once again hit breaking points. California, with more than nine thousand fires, saw the first ever wintertime megafires during "what should be the peak of the state's rainy season."[3] As one observer of these fires noted, "I was expecting to see snow on that mountain, and now the thing is on fire."[4] The notion of well-defined fire seasons is approaching obsolescence. Fire managers are "moving away from calling them 'fire seasons' to just calling them 'fire years' because there's not a season in particular. It's just year-round . . . and 2017 and 2018 can attest to that," as a fire manager from the southwestern United States told me.[5] In mid-November of 2018, the Camp Fire struck Paradise, California, claiming at least eighty-six lives and becoming the deadliest fire in the state's history. In mid-September of 2020, entire towns went up in flames in California, Oregon, and Washington in the worst fire season on record.[6]

In what follows, I explore how wildfire professionals and emergency responders in the western United States contend with fire behaviors that often test preestablished plans for control. As the geographic footprint of extreme wildfire expands, I focus on the south and northwestern United States, where a second-generation fire manager told me he needed "a bigger map" for the growing area that fires have consumed. With simultaneous large wildfires burning in 2017, there "weren't enough people to put them out," he said. As happened in California in 2020, one fire had tripled in size overnight, when lower temperatures and higher humidity would typically cause flames to die down; the fear was that it would merge with other fires to become a megafire.

In many parts of the US West, forests and landscapes have evolved with controlled ground fires as well as uncontrolled burns.[7] In midsummer, the fire manager and his team of firefighters sketched lines on a map to demarcate containment perimeters around another fire, relatively small at the time. Based on topographic patterns and weather conditions, the perimeters represented a best guess of how far the fire could spread, and they guided fire-suppression decisions (for "holding the fire"). Firefighters would allow the fire to burn up to one perimeter, and they would actively "burn out" areas along another. Getting rid of combustible biomass (the "fuel load"), they could redirect the fire, preventing it from reaching communities. By late summer, fire filled the area, subverting plans to keep it from spreading further ("holding the line"). With behavior that was typical for that season, one of the fire's flanks had beat the firefighters to a critical perimeter.

Once it looked as though it was dying down, firefighters went to assess whether they could engage the fire there. They got caught in a "blowup." Winds rapidly shifted, and the spot fires that surrounded them became a high-intensity burn. "It didn't just walk," as one witness said. "It torched out every tree it hit, burned down to the rock." The firefighters were overwhelmed by heavy smoke and heat and started deploying their emergency fire shelters. These are portable four-pound devices of last resort carried by all US Forest Service firefighters. In a seemingly miraculous instant, as it was consuming acre after acre, the fire just lifted up, as if it took a breath, briefly clearing the air and giving the firefighters an opening to run through to get to safety. Even when crews escape being overtaken by a fire like this, their ordeal counts as a fire "entrapment." Evidence from such close calls (which can include partially deployed shelters, packs of equipment, and the various, sometimes melted, tools that are left behind) becomes part of a formal investigation into the circumstances of narrow escape, as well as a "lesson learned."

A so-called flash drought, or an abnormally rapid onset of drought, made the wildfire season hardly operable—the worst since the 1910 Big Burn, when separate fires merged into a devastating inferno that burned across Montana, Idaho, Washington, and British Columbia.[8] The firefighters were lucky. No one died. An incident commander, one of only sixteen in the United States assigned to manage large or complex disasters, such as wildfires, told me that with multiple fires burning at once, the dreaded wind events that could have made the fires merge with other fires across the landscape never occurred, sparing surrounding communities.[9] The commander and his team played a critical role in engineering what he called "soft successes," when resources couldn't be concentrated. In one fire, instead of "lining up crews nose-to-tail to go up the mountain on this big, direct effort," the strategy was to work indirectly, backing off from the fire until personnel could bring what eventually became a low-intensity fire "down a hill and under our terms to a place where we could be more effective holding it against the line." Both experience and luck favored the commander this time.

Over their lifetimes, firefighters have seen dramatic changes to the definition of a normal workload. One firefighter began fighting fires in 1996 in the Flathead National Forest, working on a district crew to pay for college. Of the three months he spent with the crew, he estimated having devoted 10 percent of that time to fighting fires, "maybe less," and those consisted of small fires, some the size of the room we were speaking in (about a hundred square feet). He said one would have been "lucky back then to get on a big fire."

Smaller fires are usually subject to "initial attack" to prevent them from spreading.[10] But this has meant that ecologically appropriate and necessary fires do not burn. We enter into the so-called wildfire paradox, or the idea that the more wildfires are suppressed, the worse the wildfire problem becomes. With roughly thirty-two thousand people qualified to fight wildfires in the United States, the pool is limited and workloads have increased: the paradox is too costly to sustain.[11]

But a fire manager for the Santa Fe National Forest, who told me he had "tapped everyone who was available at some time or another" in recent years, had also seen something all too rare: a lightning-caused wildfire that was allowed to roam around. It was a low-intensity burn, helping to decrease fuel loads and maintain a landscape that can moderate wildfire severity and spread. The burn was emblematic of "a desired future condition where wildfires can burn through the Santa Fe [National Forest] without the Forest Service trying to put them out," as one supporter of such burns noted.[12] The fire manager described the fire almost like a living creature, curiously wandering about, and just needing some monitoring as it ran its course. Fire scars from previous burns in the area reduced its flames to creeping tendrils that slowly coiled through mosaics of rocky mesas and drainages.

The anatomy of the fire was distinct. It never had heading flames, bending and moving in the wind's direction and "demolishing everything in its path." The landscape kept the accelerating effects of wind in check. In its search for fuel, the gentle entity would "take a finger here and go out, the rain would put it out. It would run out of fuels and then go in a new direction." He saw the fire as "burning naturally, the way it's supposed to when man is not involved, trying to suppress it just because somebody in town doesn't like the smoke." In fact, surrounding communities "didn't know that there was an actual fire going on." There was very little smoke, "and if there was, and unless people smelled it, it would just blend in with the clouds."

One afternoon in June 2018, as dust storms from the prolonged dryness encircled parts of northern New Mexico, I met a fuels specialist I call Rey Mason at a ranger district station of the Santa Fe National Forest. Mason had been conducting prescribed burning in the area; I had read about his most recent plan and wondered why it had been canceled. Mason is a third-generation firefighter. His father worked for the Forest Service, first as a lookout and later, after retiring, getting involved in timber sales. His grandfather worked as what would now be called a seasonal firefighter, coordinating thinning projects in northwestern New Mexico. Mason explained that prescribed burning typically happens in the spring and late fall, when favorable conditions—cooler temperatures and higher moisture levels—prevail. A prescribed burn intended to mimic the restorative effects of the lightning-caused ignition had been planned for the Monday "after the Easter holiday." The burn never happened because the window for safely conducting it closed in just three days. Melting snow from a mesa made the area too muddy on the preceding Friday. By Monday, the soil was already too dry.

Mason had spent months planning the burn, getting the permissions and creating a fire plan based on what he projected to be the characteristics of the fuel, such as vegetation moisture levels, for that particular window of time. When he measured the moisture levels, "it was just unbelievable how stressed the fuels were" because of the drought. So prescribed burns were called off.

At first, he told me that patterns were normal. "In my eyes, it's cyclic," he said. But the snow had barely melted and the fuels were already drought stressed. A few minutes into our conversation, he changed his mind: "I've never seen this. Nobody has ever seen this. You talk to folks who are 85, 90 years old and they've never seen these kinds of drought conditions. My

105-year-old aunt, a lifelong resident of New Mexico who, coincidently, we just buried last week, said she had never seen this type of drought. So that tells you the history right there."

"The Upper Limits of Bad"

With each year, for wildfire professionals trying to plan for adequate suppression, recruit personnel, or carry out prescribed burns, the ground is shifting. This shifting ground reflects larger patterns. One key study found that the global area burned by wildfires doubled over the past three decades. As Matt Jolly, its lead author, told me, wildfires are changing from one type of fire to another, much faster type—for example, ground fires are transitioning faster to crown fires, in which fire moves up and spreads rapidly from treetop to treetop. Drier conditions redefine fire potential; instead of a gradual buildup, a "switch is thrown."[13] In that moment, "no matter what you do, you're not actually going to effect change." He referred to this state as the "the upper limits of bad." Here, assumptions about what combusts—dead plant biomass versus live plants (whose burning differs from dead woody fuels and is less understood)—do not always hold. The borders between rural wildland fires and urban conflagrations that fires trample over are "not relevant to the physics of what is actually happening." Or, as the fire historian Stephen Pyne observed after a sudden firestorm destroyed city blocks in Santa Rosa, California, the fires "seem to be going where the houses are.[14]

As fire managers, fuels specialists, and firefighters find themselves in the upper limits of bad, they will have to define criteria for inoperability more explicitly, and firefighting organizations will need to recalibrate protocols of response to conditions that are rapidly changing. As Bill Armstrong, a forester and fuels specialist with the Santa Fe National Forest, told me, "We should be treating these wildfires the same way we do hurricanes: get the hell out of the goddamn way." Meeting Armstrong in 2015 (before he retired) helped me construct a path of inquiry through regime shifts that took me to, among others, wildfire scientists at the Fire Sciences Laboratory in Missoula, Montana. They were contending with future managements of starkly compromised forest ecosystems without the "surrender [of] the capacity to think," in the words of feminist scholar and historian of science Donna Haraway, who cautions against any such surrender in a time of "onrushing disasters, whose unpredictable specificities are foolishly taken as unknowability itself."[15] The fire scientists agree.

During his years of forestry work, Armstrong took up wildland firefighting, but mainly focused on prescribed burn treatments in the Santa Fe National Forest. He is also a critic of the US Forest Service's century-long policy of fire suppression. In a local forest supervisor's office in Santa Fe, New Mexico, he described a decade that saw an "exponential rise in area burned." One particular fire stood out: the 1996 Dome Fire in northern New Mexico, whose physical behavior was unlike anything he had ever encountered. In his words, "In the eighties, when I started, intense fires were anomalies. What we thought was a freak incident became a wake-up call that nobody woke up to. It was a plume-dominated fire—more like a firestorm, with so much energy released in such a small period of time. We just weren't expecting that kind of fire behavior."

Plume-dominated fires are characterized by huge updrafts of hot air pulling burning embers and billowing smoke into cloud-like formations. They have become more common over the past few decades as drier fuels make for higher-intensity fires that lead to greater interactive "coupling" between a wildfire and atmospheric conditions. Armstrong described the coupling in vivid terms: "As material moves up through the clouds, it cools off, and when it cools off, the weight of the cloud can no longer sustain itself. Then the clouds collapse, and when they collapse, they throw shit everywhere." One fire can play out this sequence of events repeatedly.

By the first decade of this century, plume-dominated fires had become too familiar. Armstrong was a first responder to the 2011 Las Conchas Fire, which, like the Dome Fire, defied behavioral models and assessments. Wildfires usually move with winds, but Armstrong noted that the plume-dominated fire at Las Conchas "burned with greatest intensities and rates of spread against the wind. It actually managed to push its way up against the wind." This meant that the fire, "with its own internal winds and its own weather system, . . . was feeding itself."

Unlike previous fires, the Las Conchas Fire eerily and aggressively "burned into areas where we thought there was absolutely nothing left to burn. It burned through the Cerro Grande wildfire scar of 2000.[16] It burned through the old Dome Fire scar of 1996. We could not have predicted that kind of fire behavior." Scarred landscapes, in which fires ripped into old fire scars, can take a long time to heal. In the meantime, they suggest a breach in once-secure models of wildfire behavior, opening a gap between what is expected and what is encountered in the field that, in Armstrong's mind, is getting too wide to bridge. The gap has other effects, at times foiling emergency responders' ability to follow a protocol of knowing "where to 'cut line,' or create fuel breaks, how big the lines should be, and when to run."[17] For him, discrepancies between what analysts can predict and what emergency workers face reflect a new pathological normal in which, sometimes, there is no

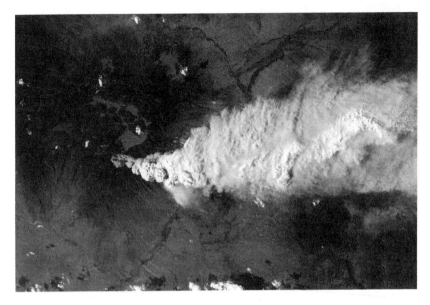

FIGURE 5.1. The 2011 Las Conchas Fire in the Santa Fe National Forest, as viewed from the International Space Station (photo by Los Alamos National Laboratory, licensed under CC BY-NC-ND, https://www.flickr.com/photos/losalamosnatlab /5926991831).

justification for fighting fire. People, including firefighters, just need to "get the hell out of the way," as they do with hurricanes (presuming they can).

Fire Hurricanes

At first, I did not know what to do with Armstrong's wildfire/ hurricane comparison, but it kept nagging at me. Both calamities, I came to realize, are akin to what Timothy Morton calls "hyperobjects." They are "massively distributed in time and space relative to humans," in some cases even comparably so.[18] Both feed on gradients in temperature, wind speed, pressure, and humidity, among other variables, and the carbon emissions driving climate change let them gather the energy to achieve intensities that can never be reversed.[19]

Yet they are subject to very different forms of physical management. In Armstrong's absurdist image of a hypothetical disaster: "We don't get out in front of hurricanes with fans, trying to change their direction. We don't get out in front of tornadoes, trying to turn them around." While hurricanes and tornadoes are allowed to run their course, wildfires are suppressed (more or less successfully). When I spoke with emergency coastal managers just an hour from my hometown in central New Jersey, I identified other distinctions. Unlike fire management, emergency coastal management is a high-tech affair. The decision to evacuate coastal populations is based on a relatively clear and automated signal. As one manager in Monmouth County described it, it is a "threshold based on the National Weather Service's forecasting of tidal surge or tides based off the Sandy Hook [New Jersey] gauge. . . . So, just throwing a number out, if the gauge reads nine and a half feet above mean low or low water at the Sandy Hook gauge, that triggers my voluntary evacuation of everyone in Monmouth County in Zone A." In wildfire management, there is no such single or automatic threshold to signal retreat. From an operational standpoint, it can also be difficult to know exactly when to step back. As one specialist suggested, "I don't think using the term 'retreat' is very popular with firefighters. Retreat is seen as a retasking." He told me about an "alarm-bells-going-off-and-you're-[still]-going sort of mentality," in which "there is no mechanism to say no."

These hyperobject comparisons and distinctions led me to other awkward facts. Even as billions of dollars are being poured into wildfire suppression to protect public lands and private property, the efficacy of suppression is "unknowable in an exact sense."[20] What would have to happen, I wondered, for retreat to become an option for responding to wildfires? What changes to current practices and paradigms would have to be made?

In following the path that Armstrong set me on, I met fire researchers who, in some instances, are starting from scratch in their attempt to model wildfire spread. I encountered supervisors who, facing abrupt shifts, are now urging workers to shed old mental slides, or once-informative guides to fire behavior, that can now lead to serious errors. As we will see, the shedding is too slow for environments that are rapidly switching. These committed experts are all too aware of the costs of delaying interventions that could help turn landscapes into fire-adapted mosaics (or patches of vegetation that can inhibit fire spread). As a midcareer supervisor put it, "It's hard to capture and present these costs to a public that has an expectation of us to protect their homes and protect everything." And that's where the challenges are. Determining what should replace the old slides is an ethical as much as technical question about whether we can all see the same fire.

To appreciate some of these challenges, consider one aspect of the devastating 2018 Camp Fire. An initial news report described the fire as a "wildfire [that] is hopscotching across a northern California county at a rate of roughly 80 football fields per minute, forcing evacuations, injuring residents and firefighters, and sending families racing from their homes."[21] The images of a fast-moving fire were indeed apocalyptic. But as I read the story, I tried imagining what "80 football fields per minute" actually meant. Was it a real calculation of the wildfire's growth rate?

I emailed Jack Cohen, renowned for his research on wildfire combustion and home protection in the wildland-urban interface where wildland and human development meet. Cohen worked for the US Forest Service Rocky Mountain Research Station's Fire, Fuel, and Smoke Science program until his retirement in 2016 (his research is the subject of chapter 8). I asked him about the article's claim about the fire's growth rate, and he told me that it is always relative to some baseline. If fire size was assessed in, say, the first minutes after ignition, that type of growth rate would be huge and unrealistic. Assessed once the fire size reached, say, ninety thousand acres, that growth rate "would be largely just an active fire day across a large perimeter." Add continuous forest, drought conditions, and strong winds and that would be expected, not unexpected, fire behavior.

In the back-and-forth email exchange, Cohen was pushing back on the article's "tsunami of flames" image. He thought its tone precluded a real parsing out of what made the town of Paradise burn.

I emailed him some of the photos I found online, including one showing complete destruction of a large tract of housing in the town—evidence, I thought, of the tsunami.

He encouraged me to look more closely at the image. Look, he said, all the structures are destroyed. But the conifer tree canopies, adjacent to the destroyed structures, are mostly unconsumed (and, in fact, green). So did a massive, high-intensity thermal exposure consume Paradise? According to Cohen, the green trees suggest that the answer is no. Had such an exposure occurred, he said, it would have consumed them too.

What, then, incinerated the structures? The wildfire lofted firebrands as far as half a mile away, but the firebrands that flew from low-intensity ground fires or from one structure to the next were more damaging, such that, according to Cohen, "the community largely burned independently of the wildfire."[22]

And if it wasn't just wildfire, but infrastructural choices (around building design and development in fire-dependent ecologies, and indeed, fossil fuel energy use) that caused destruction, then what is the true scope of the emergency? The ignition patterns the fire scientist asked me to observe highlight the challenges of intervention as more communities face devastation. Those patterns also suggest that when certain notions of crisis are magnified, certain images of inevitability can lock into the imagination. But what if the only slide imaginable wasn't of dystopia, but of future life-worlds that are livable and ecologically compatible as matters of political design? Surely this is a more achievable prospect than a war on fire? Indigenous counternarratives illuminate the stakes of this crossroad. In countering misguided policies and histories that make wildfire ever more catastrophic, they point to other futures that are already underway.

6

"You Can't Take Fire Away"

Is the problem of wildfire unprecedented, different from what it was before? Fire requires heat, fuel, and an oxidizing agent (typically oxygen) to ignite: its physics are the same everywhere. Yet to the extent that any of these elements can be added or subtracted, fire can be made to behave differently, even in a "new" way. In the United States, one can trace changes in fire behavior, in part, to a practice of fire "exclusion" that was applied "to all forests without regard to a context of place."[1] This practice also includes a history of violence involving the decimation of Native American populations, whose fire-management strategies of putting "fire on the land" yielded remarkable ecological stability.[2] Studies of the impacts of colonization between 1640 and 1900 show just how consequential these strategies were (and remain) for reducing the risks of destructive wildfire. After European contact, when fire started to be excluded, forests grew rapidly, triggering "landscape-scale fire events" and "an increase in the frequency of extensive surface fires."[3]

In other words, it wasn't just suppression, but the elimination of the knowledge and practices of Indigenous people that created conditions that today bring about more intense wildfires. To call them unprecedented, or apocalyptic, obscures how fire exclusion becomes a euphemism for far-reaching settler colonial occupations that erase Indigenous histories and limit circumstances for future adaptation.[4] In the twentieth century, large, fire-resistant trees, which had been made so by long-standing Indigenous burning practices, were logged for commercial purposes, only to be replaced with "widespread continuous forests with, on average, smaller trees and much greater fuel loads."[5]

Exclusion—be it in the name of fire suppression, settler colonial violence, or environmentally destructive commercial extraction (logging)—informs a long trajectory of regime shifting in landscapes that are now ready to ignite. The option of controlling wildfires that are now supercharged by climate change, let alone fighting them, is diminishing. But they are not a product of a rupture, as if the fires we see today are too novel or alien to comprehend. Framing them as such furthers Indigenous dispossession and obscures basic truths about the evolutions of wildfire under colonialism and capitalism. As an Indigenous fire manager in northwestern Montana told me, "The tribes have always said, you can't take fire away."

In his words, futurity is at stake—but not the dystopic kind "from which others must be saved," as scholar-activist Kyle Powys Whyte, enrolled member of the Citizen Potawatomi Nation, asserts. Building on Mark Rifkin's idea of settler time, in which Indigenous groups are assumed to be one among other groups in a purportedly neutral present, Whyte introduces a distinction between settler time and what he calls Indigenous time.[6] The latter is about the anthropogenic infliction of extinction on particular groups that, having lived through iterations of past extinctions, can remake ideas of the future, as Laura Harjo writes, and figure out "how to get there." In this wayfinding work, Indigenous time connects "unactivated possibilities" with a chosen future that "enables a community to propagate."[7]

Settler time, on the other hand, is a different creature. It unleashes extinction, yet it does not know where it stands in relation to what it has unleashed; it has no way of negotiating the future except in a "dystopic" mode. In such a mode, the future becomes a shrinking resource; human influence narrows to short-term or last-ditch emergency responses that, in a form of recolonization, tend to exact greater tolls on Indigenous communities, particularly those making up the "frontline communities" of environmental devastation.[8] Whyte rejects the "urgency" of climate change on the grounds that in settler time, "saving what for many of our people is a dystopia is not a very good strategy for allyship . . . because we're trying to get to another point."[9]

Getting to that point runs through fire not just as a natural entity but as a focal technology for Indigenous land management and political self-determination. The Confederated Salish and Kootenai Tribes of the Flathead Nation used fire as a powerful means of modulating wildfire behaviors in the complex mountain terrains of the Northern Rockies.[10] *Sxʷpaám*, a word meaning Makes Fire or Fire Setter, refers to a tribal person responsible for starting fire and who held knowledge of its effects on animals, landscapes, and plants, "inherited from generations of *Sxʷpaám* before him."[11] A slightly different translation of *Sxʷpaám* is "someone who makes fire, here and there, again and again," reflecting an act of continuous cultivation of or tending to fire-adapted habitats and species. As Germaine White, the former information and education specialist for the Confederated Salish and Kootenai Tribes, pointed out in a 2018 conversation, the "reduplicative and distributive" nature of such activity is built right into the name.

More than mere manipulation of heat, fuel, and oxygen, this relationship to fire involves "observation, experimentation and spiritual interaction with the natural world" and engagement with what White calls a different operating system. If one acts with a sense of responsibility for managing this natural world, she affirmed, the system confers well-being "for all future generations yet to be born." To ensure its reduplicative and distributive nature, something of greater or equal value is given in return. Fire exclusion in settler time has disrupted the reciprocal quality of the system, but its presence as an extraordinary gift remains intact.[12]

White directed the Fire History Project (2005), which supports education and stewardship efforts toward reclaiming "the gift of fire" on lands where fire has been excluded.[13] The Confederated Salish and Kootenai Tribes (CSKT) include the Bitterroot Salish, Kootenai, and Pend d'Oreille Tribes. With the Indian Self-Determination Act (1975) and later the Tribal Self-Governance Act (1994), the CSKT and its Forestry Department took control of their land-management decisions. A comprehensive Forest Management Plan laid out a goal of reintroducing fire; it was approved in 2000, a year marked by destructive wildfires in the western United States.

A few years later, the Fire History Project was born. As a unique multimedia learning tool, the project interviewed Tribal elders about the impacts of putting fire on the land across an array of ecosystems and vegetation types. With fire, their parents and grandparents could carve out mixes of prairies and forests, some like "open cathedrals of ancient ponderosa pine."[14] They maintained diverse plants and animals and built forest microhabitats or "mosaics" whose variable microclimates can slow or stop a fire. They cultivated plant species whose growth or regeneration relies on low-intensity fire to open up seedbeds and assist in the uptake of nutrients. What are now called "fire-dependent" landscapes are, in fact, the product of over ten thousand years of fire use across ancestral territories that include western Montana and parts of Idaho, British Columbia, and Wyoming.[15] Practices of this kind can today help protect "healthy plant communities, clean air, water, undisturbed spiritual sites, prehistoric and historic campsites, dwellings, burial grounds, and other cultural sites because these areas reaffirm the presence of our ancestors." As Joe Durglo, former Tribal chairman of the CSKT wrote in the Tribes' 2013 *Climate Change Strategic Plan*, "Our survival is woven together with the land."[16]

Salish place-names speak directly to this sense of interconnectivity, linking particular mountains, hills, clearings, lakes, and rivers with Indigenous uses or historical happenings. The Salish-Pend d'Oreille Culture Committee (now the Séliš-Qlispé Culture Committee) protects and perpetuates the Tribes' histories, cultures, and languages. As a part of this mission, it has documented thousands of place-names across Séliš (Salish or "Flathead") and Qlispé (upper Kalispel or "Pend d'Oreille") homelands, based on Tribal elder knowledge and archival research. Members have produced a landmark place-names project called the Séliš-Qlispé (Salish-Kalispel) Ethnogeography Project, a compendium that will culminate in a multivolume atlas, Skʷskʷstúlexʷ: Names Upon the Land. Their work is part of global efforts to confront settler-colonial erasure and to restore place names to Indigenous homelands, be they in the United States, Canada, Australia, and elsewhere.[17]

Some place-names, such as Scattered Trees Growing on Open Ground, reflect microhabitats that regular burnings shaped to modulate fire's severity, or they signal a transition from a certain forest to an open prairie, as in Coming to the Edge of the Forest. At a place called It Has Camas, a sacred fire-dependent plant appeared annually or after a good fire (there were "so many that they appeared to be shimmering blue lakes").[18] According to the Fire History Project, "place-names are of great importance to our people, because they often record information about the cultural ecology—how the land was used and managed." This is especially true of the "place-names [that] tell us where fire was used in beneficial ways to maximize plant and animal resources."[19] Returning to the notion of Indigenous time, such benefit accrues in relation to an operating system in which human and nonhuman life could "communicate seamlessly," in White's words, working toward a kind of stabilization that is coordinated by both.

In other words, within the operating system, stability isn't an abstraction; it consists of a real set of relations among human, nonhuman, and more-than-human beings. As Candis Callison writes, a "continual assessment of the state of these relations" creates and maintains stability.[20] Métis anthropologist Zoe Todd points out that such assessment also involves everyday "acts of tending to, enlivening, and mobilizing relationships." Building on Vanessa Watts's notion of Place-Thought, Todd notes how, in Place-Thought, relations among human, nonhuman, and more-than-human beings are constantly enacted in a shared time and space.[21] In writing that "our survival is woven together with the land," Joe Durglo echoes Watts's idea that "[o]nly if the land decides to stop speaking to us will we enter the world of dislocation where agency is lost."[22]

The settler colonial approach of fire suppression has tried to dismantle this continually assessed operating system, relegating the problem of stabilization to specialized management sectors and scientific silos. In settler time, the interdependent practices that make up stability (the assessments and relations on which they are based, in Callison's sense) remain marginalized, while those responsible for over half of the world's cumulative emissions (the wealthiest 10 percent of humanity, including the United States, emitting more CO_2 than any other country to date)[23] can be surprised by the apparent suddenness of their deleterious impacts.

As Indigenous scholars have argued, forms of Indigenous land stewardship are not a supplement to dominant settler practices; arguably, they set the standard for stabilization and, given their measurable impacts, can outperform them.[24] Tree rings sampled from four-hundred-year-old dead and living ponderosa pines from the Santa Fe National Forest suggest Pueblo people using fire with confidence—so much so that they "simultaneously add[ed] more ignitions resulting in many small-extent fires."[25] In western Montana, stumps of centuries-old ponderosa pines show that lightning-caused fires burned them once every ten to fifteen years. In some areas, in fact, they were subjected to even more burning (once every five to seven years). Relatively recent added ignitions made fire burn at lower heights and intensities, so much so that pines would "barely feel a thing."[26]

One can listen to Felicite "Jim" Sapiel McDonald (1922–2017) who was in her early eighties when the Fire History Project asked her about high-frequency, low-intensity fire and what it did for the land.[27] Burned with a fire that wasn't too hot, it would enhance huckleberry growth. At other temperatures, and just like pruning, it helped regulate certain plant heights. The Fire Setter knew "when to light, how to light, where to light" to clear sites that were tangled up in overgrowth, open trails, or fireproof camps.[28] Making fire "here and there, again and again" boosted chances of a good hunt, increased fodder for game animals, produced fertilizing ash, and generated plant foods and medicines, such as the camas plant, among many others.

For Robin Wall Kimmerer, a plant ecologist and enrolled member of the Citizen Potawatomi Nation, fire became a luminous instrument. She writes, "The fire stick was like a paintbrush on the landscape. Touch it here in a small dab and you've made a green meadow for elk; a light scatter there burns off the brush so the oaks make more acorns. Stipple it under the canopy and it thins the stand to prevent catastrophic fire. Draw the firebrush along the creek and the next spring it's a thick stand of yellow willows. A wash over a grassy meadow turns it blue with camas. To make blueberries, let the paint dry for a few years and repeat. Our people were given the responsibility to use fire to make things beautiful and productive—it was our art and our science."[29]

It also made the "landscape that European-Americans first saw when they traveled west," writes White; the landscape wasn't a "natural terrain in the sense of being untouched by humans.[30] As a former fire manager told me, "when Europeans came this way, they started seeing, especially in the summer months, a lot of smoke. I mean, there was a lot of smoke all the time and they talk about it in their writings." Tribes were setting fire to the land "to clean it up" and to produce "a park-like effect."

Today, fires are tackled with bulldozers, chain saws, and chemical flame retardants—an approach that has proved less than optimal. The fire manager was talking about restoring not a lost paradise, but a world where method and circumstances align. To return to tree rings for a picture of alignment, this time from the Jemez Mountains of New Mexico: archaeologists found that no villages of the Towa-speaking Pueblo people burned in wildfires before 1680, the year of the Pueblo Revolt. Sometimes, the fire stick even "painted" the climate itself—producing a large-scale effect overriding year-to-year climate variation.[31]

Here, then, is evidence of how Indigenous fine-scale practices could achieve stabilization within dynamic environments that include breathtaking mountains rising almost ten thousand feet from the Mission Valley floor; these mountains held new fire scars that were overlaid on old ones, creating vital edges and patches for fire to move. Here, the Salish welcomed the lightning that caused fires: as fire iterated and reduplicated as an operative system, it did so in ways that also made them and other Indigenous groups into effective geoengineers.

When traditional burning ended, the life and resources that Indigenous "biophysical stewardship" mediated and kept in states of growth faded.[32] However, there is a distinction to be made between loss and stagnation. When Louis Adams (1933–2016), a Salish tribal elder and cultural leader, was asked to offer observations on climate change patterns for the CSKT *Climate Change Strategic Plan*, he pointed to the need to pay attention to nonhuman worlds in ways that could protect life: "That's what the old people said, if you see an eagle flying around, a hawk sitting on a tree or a meadowlark sitting on a post, rabbits coming around close to you or any of these little creatures that come fairly close, they are telling you in their own silent way 'hey,' we are still here, we were here when you got here and we will be here with you till the end and that's why you are supposed to take care of them and that's why they check on you once in a while, because they have no voice and that's what I have in my heart."[33]

In 1891, the federal government forcibly removed the Salish from their ancestral Bitterroot Valley homeland to the

Flathead Reservation, where they joined the Kootenai and Pend d'Oreille Tribes. Chief Charlo led the Bitterroot Salish at the time when their lands were seized. In an 1876 speech on "Indian taxation" printed in local newspapers, Charlo commented on the vandalism of white settlers and squatters: "His laws never gave us a blade of grass nor a tree nor a duck nor a grouse nor a trout. . . . You know that he comes as long as he lives, and takes more and more, and dirties what he leaves."[34] In 1875, Pend d'Oreille tribal members were hunting near the Canadian border when two members who were setting fire to prairie grass were shot and killed by so called officers of the international line. The lethal repression of Indigenous people for burning is part of a larger history of fire's elimination in which families had someone who could be shot and killed for just doing what they do.[35]

With invasion, land theft, and settlement came a cessation of stewardship over fire-dependent mosaics of the kind that could override certain climate influences. Congested and fire-prone landscapes quickly evolved into other kinds of inflicted damage. The landscapes that evolved were altogether different from those with carefully made "park-like" effects. Where overlaid fire scars once held fire to a dependable logic, now a vicious cycle of choked trees and suppression-produced fires took over, making a "mess" as the fires themselves became catastrophic, and constraining options for the Tribes' future adaptation.

By the early twentieth century, the commercial imperative of protecting valuable timber drove fire eradication and suppression policies. Douglas-fir trees started crowding out remaining stands of old-growth Ponderosa pine trees. Timber theft and land claim fraud abounded. The repression of a dominant way of life in the region intensified, as did the pyrogeographic consequences of white encroachment. In 1908, in what is known as the Swan Valley Massacre, four members of a tribal family hunting party entering traditional hunting grounds along the border of the Flathead Reservation were shot to death by a state game warden.[36] In 1910, the rogue sparks of coal locomotives bringing waves of white settlers helped start a disastrous firestorm across the western United States. That firestorm, the Big Burn, fortified the case for a well-funded federal Forest Service, which also lobbied Congress for legislation to "emplace sustainable forest management on the new national forests."[37]

Fending off repeat disaster was a narrative that sold itself. The number of rangers, lookouts, and firefighter crews expanded, as did the railroad that moved the region's timber, minerals, and agricultural goods to national and international markets. There would be no tolerance of fires that scarred large trees and reduced their monetary value. By 1935, all reported forest fires were to be contained, controlled, or put out by 10 a.m. (known as "the 10 a.m. rule," which is no longer in force). After World War II, the all-out assault on fire was waged with military surplus vehicles and staple armaments that included "firefighting planes [and] an expanded road system for sending in fire trucks."[38]

Over time, the war strengthened the enemy. Non-Indigenous firefighters from the Northwest and the Southwest in their forties talked to me about how wildfire size has changed in their professional lifetime. An assistant fire manager working for a neighboring national forest got his start on an engine crew and then spent twelve years working on hotshot crews (or mobile teams of highly skilled wildland firefighters). He told me, "Our understanding has not been able to keep pace with the changing environment and the complexities that we're seeing. And that's the crux, with most people, especially folks who have risen to senior leadership positions, their slides don't match up with reality" (on such slides, see chapter 5). "For the vast majority of our history," he said, we believed that "[fire] was a phenomenon that could be controlled." Today, however, "we're never winning." He spoke to me about the difficulty of "getting ahead of fire" and getting "wrapped around the axle" with fire. Some of his colleagues still cling to the same slides, saying, "We just need a few more firefighters or we just need a couple more air tankers. They don't say 'this isn't possible, that it just isn't in the wheelhouse.'" Meanwhile, at the Flathead, a 2007 megafire would destroy an area that held tribal significance because of its vegetation and remaining Ponderosa pine—"and an area for cultural use was lost." Place with Many Large Trees and Brush Here and There had been sculpted by fire behaviors known to occur reliably across millennia: "Right now, it's just a big fire trap."[39]

With these shifts, coordinating in the face of disaster remains a central goal. CSKT wildland firefighters work with agencies in multiple state and local jurisdictions to manage and suppress fires. Ronan, Montana, is home to one of the administrative centers of the CSKT Division of Fire. There I met Tony Harwood who, among others, has played an active role in prescribed fire, forest planning, and fire management for over three decades.[40] He also gave lectures about place naming and Indigenous uses of fire. A fire scientist at the Fire Sciences Laboratory in nearby Missoula, Montana, encouraged me to speak with him because he believed that Indigenous fire practices set a standard for non-Indigenous fire management. I also met Ron Swaney, a third-generation firefighter and current CSKT fire manager, and Bob McCrea, a former CSKT fire manager who had fought fires since 1966. McCrea had been a smokejumper[41] and was a wildland operations specialist for the tribes. All three stressed the importance of interagency partnerships in the mission of controlling fire. McCrea, for example, trained "Type 1" incident commanders in the national wildfire incident command structure, while also putting fire on the land. Knowledge and resource sharing are critical. "No one agency can do it alone," McCrea said, as we were sitting in the office that had been Swaney and McCrea's strategic base of operations for worsening fire seasons.

These wildfire professionals described the alignment of different factors leading to the increased intensity and spread of wildfires in and around the Flathead, and why interagency partnerships were of utmost importance, especially as fire behaviors continue to evolve. From 1978 to 1997, the Confederated Salish and Kootenai Tribes saw only three years with large fires. Between 1998 and 2017, the numbers flipped: fire crews saw only three years *without* large fires. Fires, becoming more unpredictable, cross different jurisdictional boundaries (for example, between national forests and federal and state lands); thus interagency relationships and partnerships become even more fundamental to mitigating danger. No agency or department wants its fires to cross out of its own jurisdiction.[42] Regional mutual aid agreements also play a vital role; just as the CSKT Fire Division relies on assistance from agencies to manage its own fires, it shares its firefighters with those agencies too.

As different agencies exercise connection through common cause, the potential for working relationships will only continue to grow. The CSKT, in accordance with its 2000 Forest Management Plan, has created a model Fuel Treatment Program to achieve desired future conditions as best as possible. It has burned (or treated) over 140,000 acres on the reservation and, in some areas, fire now mimics its historical role. The differences between treated and untreated areas are stark. In some cases, large fires that would have destroyed living trees have crossed fuel-treated areas to turn into nonlethal and low-intensity surface fires. Managing future lethal fires in this way can help protect ecosystems, dwellings, and significant cultural sites, as well as make the fires themselves safer to engage.

The CSKT fuels program works with its partners to meet these and other objectives and to share knowledge. In the midst of fire, the work of creating and safeguarding desired future conditions is constant.[43] Ron Swaney looked out his office window, which faces the resplendent Mission Mountains that surround Ronan—a wilderness for some (hikers and tourists), a tinderbox of dense and overgrown vegetation for others (emergency responders). As he contemplated the mountains, he commented, "People look at that hill right there, that mountain, and say, 'That's the way it should look. That's beautiful. Look at it. It's full of trees.' And that's actually not the way it's supposed to look." Its undulating surfaces are supposed to contain crosshatches with different openings, tree types, and forest ages. Recalling the mountain as it looked to his grandparents and great-grandparents, he said: "When we looked at it, we saw patches of trees. A mosaic, with foods for the animals and the people."

Reflecting on a 2007 wildfire called Chippy Creek that consumed the Tribes' sacred ponderosa pine forest, Swaney said, "It will recover, it's going to take time. The fire burned with such intensity. It's going to be a longer process to get a natural landscape recovered and reestablished. The fire took it back to the start, all the way back to lichen and moss. We'll likely never see [the forest] come back in our lifetimes . . . , back to where it was." He was referring especially to the old-growth ponderosa pine stands: four-hundred-year-old trees that grew up in fire. The sacred area that was burned "was one of the few places that family groups could do traditional practices [like gathering foods and medicines and hunting]," Harwood added. Extreme fires cross jurisdictional boundaries and burn through old fire scars that were shaped over millennia of active fire stewardship. The fires "have a certain heartbreaking aspect for tribal people." The loss, he said, is faced constantly because "there is no other place to go for us. It has been six generations since the Tribes had been removed from their lands and forcibly moved to the Flathead. This is the Tribes' homeland now."

Here, the gift of fire reappears. Camas, a six-petaled flower with purple-to-white hues, is a vitally significant food plant. It is one of the first foods to appear after a long winter, when dried food stores run low; it is welcomed back with feasts and ceremonies that start the camas-collection process. Camas has a unique relationship with fire—"it loves a hot fire." In 1994, the tribes reclaimed land and fire management from the Bureau of Indian Affairs through the Self-Governance Act. That same year, a large fire on the western side of the reservation revealed a camas collection site. The next spring, "it was just a sea of lavender inside the fire area," Harwood told me. Tribal members checked with elders to see whether this was a known camas-gathering site. In the 1940s, elders had created maps that showed it to be a historical gathering area.[44] I asked Harwood whether he was surprised by the way the camas revealed itself. No, it was more gratifying than surprising. "Well, when you think back on it, we shouldn't have been surprised about it. It was really received with joy," he said.

7

Witnessing Professionals

When a sacred plant appears in an area where wildfires have raged, elders' maps affirm it as a historical gathering site. This story holds a significant lesson about how it is possible to remain responsive to, rather than confounded by, diminished ecological futures—in this case, through an authentication of another map that is taken in with joy. An affirmation of this kind withstands catastrophic thinking that would reduce the camas's appearance to mere surprise. One of the first foods to arrive after long winters, it is received as a magnificent inheritance. The relations and knowledge that the camas's appearance holds are not "consummated by oblivion."[1] A community's joy calls out catastrophic thinking and points to horizon-acquisition as opening onto other possibilities and outcomes becoming true.

In the next pages, I want to explore elements of the converse of horizon-acquisition, the sense that the future is limited, or that its conditions will be inoperable, thus making extreme solutions seem inevitable. I call this this future-indifferent sense horizon-deprivation. Being horizon-deprived means being cut off from an ability to meet conditions where they are; or being at a loss for knowing how to even recognize the scope of loss. When the present cannot provide useful footholds for imagining what might come next, the future becomes an abstract presence; struggles against deadly wildfires appear to have no end in sight.[2] Horizon work in all its permutations is nontrivial when it comes to how firefighters think about their routines, especially when their object (fire), framed as uninterrupted spectacle, craves more resources, including more people, to fight it. This framing can create its own form of denial of collective responsibility. In an earlier chapter, I suggested the need for a new mental slide, different from a "tsunami of flames" image, and a different moral baseline from which to confront vested interests and historical legacies that, in their denial, have profited from an image of a winnable war.

In this chapter, I consider how my interlocutors weigh old moral baselines in search of new ones as they tease out personal and ethical conflicts linked to the circumstances to which they are assigned. Along the way, I explore senses of futility alongside efforts to establish a different horizon. I take my cue from Robert Jay Lifton, an American psychiatrist, who coined the term "witnessing professionals." Lifton has written about, among other things, medical complicity in war crimes during the Iraq War, the psychiatric ravages of the Vietnam War, and the Holocaust, when, as he writes, "professionals were reduced to being automatic servants of the existing regime."[3]

Lifton contrasts professionals-as-automatic-servants with the figure of the witnessing professional, a person who refuses to be reduced to or complicit with a "malignant normality," a term that he also coined. Potential witnessing professionals are everywhere. Nurses, psychiatrists, and teachers are all professionals "with special knowledge balanced by a moral baseline" as well as the "information to make judgments."[4] Calling out the unfitness of a situation, witnessing professionals prevent indifference from taking hold in their institutions; they contest a malignancy that is presented as normal, even inevitable, or as a "condition into which we are born."[5]

I came to see my interlocutors and the wildland firefighters they support or depend on as a community of witnessing professionals. They work on the leading edge of a bold experiment in how collective responsibilities should be rethought when futures seem limited. Amid rapid ecosystem changes, distinctions between who is protected and who is not become more acute. Whether any one particular class of professionals should be charged with fighting malignant new normals, and often risking their safety doing so, is a key ethical question of our time.[6] The public depends on emergency responders for safety and protection. Yet every year, agencies cannot fill hundreds of orders for handcrews to dig firelines to contain and control wildfires because these crews are in such high demand.[7] Given patterns in 2020, the hottest year on record and, until then, the worst wildfire season the western United States had experienced, it is only a matter of time until environmental extremes expose the limits of this system of dependency.

Beyond a duty to protect (property, often people, and "values at risk"), my interlocutors have a central though vastly underappreciated role in protecting ecological futurity.[8] In its current form, suppression—which can involve bulldozers gouging out firebreaks in wilderness and air tankers spraying chemical flame retardants across developments, rooftops, and cars—seems at odds with this other protective role. As one fire manager told me, "When we should be doing prescribed burning, we're chasing suppression fires." He used the phrase "suppression fires" to suggest fires that are not intentionally set, as with arson, but to express his sense that suppression fires are man-made. Some see this chase as a quasi-militaristic operation.[9] Others sometimes refer to it as a more sluggish "groundpounding," a word suggesting repetitive toil of people and machines being used to "to control nature, or tip it back," in the words of one wildfire scientist. Some wildland firefighters reject the image of a selfless hero because it reinforces unrealistic expectations, which, they believe, help the public turn a blind eye to the risk and sacrifice the work entails.

In what follows, I consider the perspective of supervisors who understand the fight against wildfire to be yielding diminishing returns, and who want another path. Referring to the disposition of his "ops," or members of an "operations section" conducting tactical operations at an incident, an incident commander described to me how he could appreciate his "ops guys" being "pretty aggressive" and being "used to having everybody support them" with resources and equipment. But he knew that such a stance, involving "gearing up to continue fighting wildfire with more people and more equipment," wasn't always the right approach. As much as the ops wanted to fight fire, sometimes the fight was unwinnable, more like a "knife fight," he said, invoking an image of profound inadequacy. The idea that a wildfire should be confronted as if it were an enemy, and that this is a fight that can be won, has been a linchpin of wildland firefighting in the United States. However futile the fight may be in some cases, the actual dilemmas faced by firefighters cannot be confronted without risk to larger institutional orders, identities, and interests.

Hotshot crews are among the most experienced and elite ground crews in the US wildland firefighting force. There are over one hundred hotshot crews nationwide; each is composed of twenty people. They are a "Type 1" crew operating in the Incident Command System, a coordinated system of emergency response.[10] Hotshots lead in what is called initial attack and extended attack to contain fires, often in remote areas. They undergo grueling physical and technical training to do tactical work, often with just hand tools, such as chain saws and shovels, and very little logistical support. Originating in the late 1940s in southern California, they were called hotshots because they worked on the hottest part of wildfires.[11] Today, they are considered a national resource, shared among various agencies as a frontline firefighting force.

Superintendents lead individual hotshot crews, hiring and training crews, managing budgets and resources, and leading crews on fire assignments. They hold more qualifications than do other types of crew supervisors. As one superintendent told me, "I'm a resource for the hotshot structure, and I'm as high as I can go." Like the incident commander above him, he was concerned about the normalization of a reactive stance. He placed himself in a long line of hotshot leaders who, over four or five decades of their careers, "saw the change in the intensity and therefore severity in which wildfires have progressed."

Worst-case scenarios were simply expected, he said, and his assessment was similar to that of other superintendents: "If you talk to somebody like my old supervisor who was fighting fire in the 1980s and take something like the Yellowstone Fire (1988), that was an anomaly—that was, 'Whoa, we didn't expect this.' Now a fire gets started and, immediately, everybody says, 'Just how big is it going to get? How much is it going to cost? How many homes are we going to lose?' Then we just move into worst-case scenario mode, right away."

In some ways, forgetting past baselines takes the psycho-logical burden off the day-to-day work. Speaking on behalf of himself and the crew he assembled and trained, "this is the only type of fire behavior we have ever seen." He added that ignoring what came before helps crews to focus on the task at hand "because for us, these very large, very extreme, very long-lasting, landscape-changing, and detrimental wildland fires are the norm." Over any fire season, physically arduous duty "beats you up." The effects of fatigue, compromised immune systems, sleep deprivation, smoke inhalation, poor nutrition, and poor hygiene are cumulative, "so by the end of a fire season, you're pretty trashed."[12] More extreme fire seasons (and, in 2020 and 2021, a mass pandemic) conspire against firefighters, particularly seasonal firefighters, who are vulnerable to sickness both during and after a fire season. These workers are also facing health insurance and workers' compensation processes that are riddled with inconsistent rules. In the words of Casey Judd of the Federal Wildland Fire Services Association, "it's a mess."[13]

For this superintendent, finding new recruits who were adequately prepared or fit for demands in the field was getting harder. Hyper-readiness meant "putting everything out there" in terms of resources. But he was also skeptical of this approach. Wildland firefighting is complex business, impacting jobs and livelihoods. Myriad contractors are involved; until recently, they were "ready to blow in instantaneously to a fire-suppression site," even in very remote locations. An "entire city could be established with showers, logistics, cafeteria, game room, fuel tenders, water tenders, aircraft. . . . All these national caterers, all these planes, all these helicopters, all the port-a-johns, sinks, generators, tents, bottled water, Gatorade." That the business aspect has "taken over and become its own monster" was not lost on others up the chain of command. A former incident manager told me that "firefighting has become, on the private contractor side, such a business" that some members of the public are skeptical: "I've had people come up to me and say, 'You're just letting this burn so you can all make money, right?' Right. And it's really disheartening to hear that."

When they first started out fighting fires, both the superintendent and the incident manager recalled a spirit of selfless volunteerism. As the superintendent put it to me, "The equipment operators and the loggers in the woods would just drop what they were doing, and they'd come, and they'd bring their dozers [bulldozers]. They would just work to put this out because it was in the common interest to do that."[14] Common interest has become private interest and a multimillion-dollar industry in which private contractors compete to land Forest Service contracts to provide private ground crews; aerospace and defense companies profit from the sale of airplanes and air tankers. And, as the climate crisis grows, so too does the private firefighting industry's influence on politicians and government.[15] Suppression, in economic terms, "is a self-perpetuating beast." Once the decision to suppress a fire is made, "we're going to suppress it," as the superintendent told me—because "if this fire doesn't stay where it's at, eventually I'm going to have to answer the question, 'Why didn't you put it out when it was small? How did it get to this house? How did all this smoke pop up and affect my tourism dollars over here?'" In other words, the public's expectation that firefighters will be deployed can drive an excessive response and increase the potential for injury. As he told me, referring to his crew members, "Now that they're here, they may get injured or killed. So you see the conflict we constantly have within ourselves?"

Yet the drive toward more resource expenditure also reflects an increasing sense of powerlessness to change outcomes, especially in an active fire season. "Hitting initial fire starts hard with all we have," a resource-intensive aggressive attack, is critical to stopping fires from joining other fires and creating unstoppable megafires. This was especially true during the COVID-19 pandemic, when federal and other contracted crews were limited in working the fire lines because of social-distancing restrictions. To avoid the need for large traditional fire camps in the field, agency officials and fire managers prioritized the use of aerial equipment in initial attack fire suppression; tactical support helicopters, airtankers, and other aircraft were obtained on "call-when-needed" contracts. Greater reliance on air attack signaled a boon for the firefighting aircraft market (North America being the largest in 2020). The policy also created uncertainty for crews on the ground, at least before the worst of the 2020 wildfires started erupting. It is a well-known fact that federal wildland firefighters are underpaid, and opportunities for overtime work can provide reliable supplementary wages. But the focus on air attacks from above threatened to minimize some of those opportunities, which had some crews anxious about whether they could meet their seasonal compensation goals.[16] In other words, if commercial interests and unresolved labor issues shape the scope of suppression activity, it would follow that suppression and its efficacy is "unknowable in an exact sense."[17]

Moreover, chasing suppression fires in whatever form that takes is not an operation that treads lightly on land. One summer, while driving at an elevation of ninety-five hundred feet through a southwestern national forest, I spotted an open area in the forest, less than half the size of a football field and devoid of any vegetation. It was just black. After checking an informational website, I realized that I had stumbled upon a remnant of a suppression operation for a wildfire that had occurred a few years earlier. Its smoke had been visible some forty miles southeast of a small town in New Mexico I had been visiting. At the time, I had marveled at what is today an all-too-common sight: how smoke can change the quality of the sunset, an ashen orange pallor turning the usual combinations of orange, red, and yellow into a murky fog of deep reds and browns. I heard helicopters above the city, presumably carrying people and equipment to quell a fire sparked by a lightning strike and growing to thirty-six hundred acres, not very large by most standards.

Was it part-helipad, or was it an area that had been hurriedly cleared of potential fuels to block the fire? Stripped of brush and large trees, could it be a safety zone, where these combustible materials were diligently gouged out by bulldozers? Or was it the so-called black, an area considered protective because it is so completely burned that nothing combustible remains? If I had seen something like this before I began this work, I would have likely assumed that a part of the forest had to be destroyed in order to save the rest. But why was the forest left so degraded, literally, with a big hole in the middle of it? Was this evidence of what the wildfire community calls the wildfire (suppression) paradox, firefighting approaches that can entrench ecological incompatibility, not to mention ruin?

To understand the effects of wildfire suppression on the environment, we must consider some policies that shape decision making. Fire suppression is the Forest Service's principal activity, taking up more than half of the agency's annual budget. When a decision is made to suppress a fire as part of a fire management strategy, that decision is, for the most part, not subject to the kind of environmental review that private entities conducting activities with environmental implications (say, logging or road construction in forests) are subject to. Those reviews are governed by the National Environmental Policy Act (NEPA) and its environmental impact review and permitting process.[18] Suppression actions, on the other hand, can, at times, circumvent such requirements. The Forest Service has argued that each fire is unique (or, rather, a unique combination of the "precise location at which it originates, local weather condition[s], the condition of surrounding forest, and myriad other factors").[19] The emergency rule acts as a "wholesale waiver of all NEPA process and is applied to anything the Forest Service calls an 'emergency,' including post hoc justifications."[20] Moreover, the emergency rule often permits firefighting actions without requiring disclosure of the effects of those actions.

The public can therefore know little of the environmental impacts of suppression (or the value of any alternative practice). State laws often mirror the federal regulatory framing of emergency, so that, while wildfires may or may not be designated as "emergencies" in many states, suppression is mandated either way.[21] This decision has ecological consequences. According to one wildfire scientist, agencies "can bring bulldozers into wilderness areas; they can chainsaw and carve helipads in the middle of the wilderness and fly in helicopters; they can drop chemical flame retardant in creeks; they can put bulldozers and dozer lines [firebreaks in which mineral soils are exposed by the front blade of a bulldozer] everywhere in the wilderness—with no environmental review whatsoever."

The apparent lack of environmental review is the subject of continuing disagreement. In 2018, an ethical watchdog group named the Forest Service Employees for Environmental Ethics (FSEEE) attempted to break a cycle of harm, filing a legal brief requiring "environmental review of standard wildland firefighting tactics." In its brief, filed with the Ninth Circuit Court of Appeals, the FSEEE compared suppression logic to a dysfunctional emergency room in which a physician who "shoots from the hip, without advance planning, medical references, studies, data (and, consequently, a well-informed patient), commits malpractice and faces a high risk of bad medical outcomes."[22]

The case involved a Forest Service official allowing timber companies to mow down a fifty-mile-long, three-hundred-foot-wide swath of land, allegedly intended as a barrier (called a "community protection line") between the 2015 Wolverine Fire and nearby communities. According to the FSEEE, "The line did nothing to stop the Wolverine Fire—nor will it stop future fires. In fact, it will actually increase the risk of fire by allowing invasive plant species to gain a foothold along the line." What it did do "was provide timber—a lot of it—without the environmental reviews and public input that otherwise would have been mandatory. The line provided enough timber to fill almost 1,000 logging trucks, some of it massive old growth."[23]

While agency administrators continue to view each wildfire as a unique event, another aspect of "the emergency" rears its head: a lack of accountability that plays out in two ways. First, the practice of direct attack (an attack on fire that can include water spraying, smothering, or chemical treatments, which is not addressed by the NEPA) can produce negative consequences for fire-dependent ecosystems that are made to and required to burn, a fact that is well-known among wildfire researchers. As we have seen in earlier chapters, in fire-dependent ecosystems, fire maintains habitats that allow certain plants and animals to reproduce and grow. Well-intentioned suppression measures can leave cumulative impacts such as unburned brush to fuel subsequent suppression wildfires.

A well-known research forester saw this paradox as symptomatic of a larger problem, in which, in his opinion, "we don't have objectives for fire management." This is a striking statement, especially given the amount of resources that are being put into fighting wildfires. The Forest Service touts a 98 percent initial-attack success rate as an extraordinary feat. Citing his colleague Karen Short's research, Mark Finney told me that California fire records from the first decades of the twentieth century show the same initial-attack success rate "with horses and shovels and no communication technology [such] as we have now."

The initial-attack success rate in the early twentieth century is estimated at 97 percent. An inherited lack of fuel buildup from "millennia of active burning by Native Americans and natural fires" accounts for this high percentage. Those conditions, in this forester's view, made it "easy to have a 97 percent initial-attack success rate back in the 1910s." Today, "we have to have DC10s and 747s and every bit of technological paraphernalia" to achieve that. "California has the biggest military firefighting force in the entire world. And they still have worse fires, right? I mean, isn't anybody asking the hard question?"

A year earlier, I had been talking with a fuels specialist who had been asking the hard question about how to "treat" wildfire emergencies. Prescribed burning was at the heart of his expertise, and too much spending on suppression only took away from the prescribed burns that could be done. One too many times, he'd signed up to fight fires on "godforsaken pieces of ground, putting out fires that should have been left burning." Referring to a fire in a remote location and of a relatively small size, he said, "In two days, you had six hundred people on the site." Futility was also a running theme; they were not getting results, even "with six hundred freaking people" before the rain reduced the fire's activity to creeping and smoldering. The fire burned for several days and in an area where there were lots of dead standing trees. The situation could have been dangerous: he was concerned about the safety of the workers involved because after four or five days "that fire eats the bottoms of those trees. So, you get a lot of dead stuff that starts falling over [and you can] get a lot of people injured." In this instance, nobody got hurt, but to his mind that was fortuitous.

The activity potentially generated more exposure to risk. When it started raining in the area, the witnessing professional assumed the work would stop, leaving the rain to put out the fire. But it continued. As he told me, "It was raining on the fire. Are we getting people to stop going up there?" There was an order to put out "everything from the edge of the fire to three hundred feet in." He described a seasonal firefighter following that order as "going in, and digging, and putting it out," but it's raining. The forecast called for the monsoon to set in. "They were all there," he said. Firefighters were stringing in hose lines from reservoirs that were set up from the top of a hill, and then down into a snag patch (a grouping of burned or dead trees that can easily fall any time) "so they have water to help put these things out."

In further painting the scene, he described how, in a morning briefing, crews were told to "wind it down," meaning they were to extinguish the last remnants of the fire (also called "mopping up"): "We're getting chased out because of the afternoon's storms, lightning, and thunder. But we're still sending people down in there with the hazard of half-burned dead trees and mopping up." He was thinking to himself, "Why are we going? The fire's out. We didn't put it out. It got rained on. It's raining. Every day, it's gonna goddamn rain. The forecast shows the monsoons are in. What do you mean, 'be safe and don't do anything stupid?' Let's go home."

Crews, better-equipped as witnessing professionals with "special knowledge balanced by a moral baseline," must choose their battles in the tug-of-war between excess and futility. Amid extreme drought and heat waves, the objectives of the fight are changing. A webpage link that took me to a Forest Service description of handcrews serving as "the backbone" of a "fire-fighting army" now takes me to a page that begins with a statement about futility: wildland fires "are a force of nature that can be nearly as impossible to prevent. . . . "[24] As I will show in the next chapter, scientific research can help adjudicate directions of change, and when certain practices are likely to be effective, unrealistic, or unsafe. In facing conditions where wildfires will likely occur, or where land uses have evolved in fire-dependent ecosystems, such knowledge is essential. In the meantime, the salary for an entry-level federal wildland firefighter remains sorely low. Many of the firefighters who fought the California wildfires were state prisoners getting paid as little as two dollars a day. "Don't just cheer wildland firefighters as heroes. Give them affordable healthcare," as one op-ed implores.[25]

8

"Waiting for a Reality Response"

That wildfire in fire-prone ecologies is both inevitable and getting harder to suppress, and that fire suppression itself can heighten fire risk: these are facts that now confront the operations of emergency response. But there is a more fundamental issue with regard to the science underlying practices of suppression. As wildfire crises escalate, scientists are sometimes going back to the basics of wildfire to understand its empirical realities. A long-standing fixation on suppression, especially since World War II, meant that for decades the science of wildfire had mostly limited itself to an image of controllable wildfire. The Cold War–era experts who once dominated the relatively small field of US wildfire science created "hypothetical universes" in which fires burned only small, uniform, dead fuels on forest floors, or they reduced the unsteady forces that allow a fire to spread to mere averages. As a consequence, "false assumptions" about wildfire behavior "lingered for so long," as did an overconfidence in suppression as a mode of response.[1]

Today, the assumptions that underpinned fire suppression have real consequences for how wildfire futures are being negotiated. What we know and don't know about these futures reflects decisions about what, precisely, is "worth" knowing about them in the first place. While concerning, this pattern isn't unusual. More money is invested in understanding cancer as a biological process than in understanding how social and economic determinants subject some groups and not others to its ravages. What we know and don't know about its basic prevention and treatment is overdetermined by regulatory frameworks, public policies, and financial priorities. Knowledge vacuums are not just neutral empty spaces. They are constructed, the result of social and economic choices to prioritize and pursue certain truths at the expense of others.[2] As such, vacuums can require costly upkeep. We may know more about suppressing fires with "the biggest military firefighting force" than about the efficacy of suppression itself; more about increasing flame retardant–throwing capacity in aircraft than about how to collectively coordinate to produce more stable fire regimes.[3] Such patterns and images feed a distorted picture of inevitability: even as climate change is a foregone conclusion for the foreseeable future, the conditions that make fires burn catastrophically are not.

In what follows, I show how knowledge vacuums are created and maintained in science. I do so less for the sake of critique than to understand the struggle of enacting the research that is most relevant: if undertaken with curiosity, scientific inquiry can provide a better basis for advancing wildfire management. In other words, curiosity must be at the core of an expanded investigative framework that yields what wildfire scientist Jack Cohen calls a "reality response." At the same time, making sense of real wildfire behavior carries considerable ethical weight, especially when it comes to insuring the safety of wildland fire-fighters, who often make split-second, life-or-death calls, even when expertise is still subject to revision.

Since 1960, researchers at the US Forest Service's Fire Sciences Laboratory in Missoula, Montana, have studied wildfire behavior to build a reliable basis for making those calls. The Fire Lab faces US Highway 10, which runs northwest from Missoula, past the city's airport, and merges onto an interstate that disappears into narrow winding stretches of valley and high mountain. At the nearby Aerial Fire Depot, visitors can sometimes spot smokejumpers parachuting from aircraft into the Missoula Valley, once covered by a glacial lake, or near the Middle Clark Fork River that the Séliš-Qĺispé people named Shimmering Cold Waters (*Nmesulétkʷ*), training for upcoming fire seasons.

I am sitting with Mark Finney, a research forester who leads a fire behavior research team at the lab. His office, located near the entryway of a modest two-story building, has a clear view to Northern Rockies ranges where fires scorched more than a million acres a year before. A quiet hallway connects Finney's office to a unique multistory laboratory, where researchers have simulated and analyzed fire behaviors over six decades. There are used, abandoned, or ready-for-reuse apparatuses everywhere: wind tunnels, combustion chambers, air torches, burners, and a one-of-a-kind fire whirl simulator—all surrounded by scattered heaps of ignition sources like cut cardboard, shredded wood, matchsticks, wood blocks, toothpicks, and pine needles.

"There is an expression that everyone says," Finney tells me, "'It spreads like wildfire,' yet we don't even know how wildfires spread." Wildfire science agendas have been written largely from a suppression-focused perspective (at least in the United States), and models have not always aligned with fire's complexity. Finney and his colleagues are making up for lost time, conducting laboratory and field experiments to answer basic questions, such as how wildfires spread, and to pinpoint physical explanations for how they do, rather than should, behave.

To highlight this crucial distinction, Finney drew an analogy between wildfires and hurricanes and asked me a rhetorical question. If I had any say in the matter, would I ask people to face down a hurricane, in the hopes that they might be able to stop it or change its course? When a fuels specialist asked me this same question, I thought it absurd to act as if one could. His image of people standing before hurricanes and waving little fans underscored how disaster management can be undermined by unrealistic expectations. This time not only did the theme of futility stand out, but so did an illusion of protection.

Basic research can help to turn this state of affairs around. But, as I would learn, Finney and his colleagues are doing more than tediously filling in knowledge gaps. They are keen observers of other observers of wildfire; they know how assumptions can overdetermine what those observers see. The sense of illusion was not lost on the French philosopher of science Gaston Bachelard who, in his classic book *The Psychoanalysis of Fire*, declared that certain physical entities, like fire, can outrun the techniques used to apprehend them: "fire is no longer a reality for science."[4] In this ostensibly postscientific domain, something like combustion can look different to, say, a structural engineer trying to minimize fire's impact in an urban setting from the way it is seen by a wildfire scientist trying to make sense of its patterns in nature. (Bachelard's observers include astronomers who study star formation and even poets, whose image of fire "could clog knowledge of electricity.") Each observer, he wrote, employs modes of conjuring a related but different thing.[5]

This is no mere subjective or eye-of-the-beholder view of fire. When observers freely register or erase certain features of fire, they might not end up seeing the same fire. From the perspective of US wildfire science, wildfire's variable behavior has been one such critical but erased feature. As Kara Yedinak and colleagues write, wildfire possesses an "inherent variability that has largely eluded the fire behavior community."[6] In Finney's view, variability has been denied to such a degree that "we no longer know what fire is." To help rectify this void, a connection between observation and a more complex physical reality needs to be restored.

But why was the connection lost?[7] A part of the answer lies, he said, in the history of wildfire science which, at least in the United States, is a relatively young discipline, evolving in earnest since the 1940s. The science of urban structural fire, in contrast, dates back to the seventeenth century. The fires that once raged across European and American cities had, for the most part, been successfully reduced to "occasional and isolated threats" when combustible elements (such as thatched roofs) were banned.[8] Today, knowledge of structural materials precedes building policy (the thermal properties of those materials are so well-researched that they can be designed with flame-resistance lasting a specific amount of time). For wildfire science, Finney said, the opposite occurred: a policy of suppression took root *before* wildfire behaviors were sufficiently understood. In reducing fire's complexity, fire-behavior models served the goal of suppression so that its practices could be shown, at least in theory, to work.

In other words, relations between theory, basic science, and practical modeling were out of joint. Finney, a student and critic of this state of affairs, illustrated its conundrums by asking me if I had ever watched a dog run. It doesn't run, he said, "by putting its legs in an average position and just gliding along." But that is how some fire-behavior models posit wildfire spread. In prioritizing what they wanted fire to do over what it does, Finney's predecessors opted for the glitchy dog, "justifying how it is that fire assumes a position, and then moves." Uniform motion and steady states were favored because they offered, in the words of the early wildfire modeler Frank Albini, the "internal consistency of a well-disciplined mathematical model."[9]

Norman Maclean, in *Young Men and Fire*, describes Albini as having a "persuasive literary style that helped to make him an effective half-concealed salesman for the extended uses of mathematical models in the woods."[10] Albini spent two decades conducting defense research on problems such as weapons systems analysis, system component performance and prediction, and ballistic missile defense before joining the Fire Lab in 1973, where he worked as a fire-behavior analyst until 1985 (five years before Finney joined). Other important predecessors included Richard Rothermel, who, coming from a background in aeronautical engineering, simulated fires that had "stabilized into a quasi-steady spread condition" and "made everyone ask rate of spread questions."[11] He famously modeled fire spread as an ellipse (an image that is still used in certain instances).

Arguably, no one discerned hypothetical universes and their unsettling implications for his discipline and wildfire safety and management better than Jack Cohen, who retired from the US Forest Service in 2016 after four decades of experimental and theoretical wildland fire research. I had the privilege of interviewing Cohen several times over the course of three years, and what follows is an account of how he pursued scientific curiosity, recapturing a physical phenomenon (fire) that was losing its "reality for science."[12] He is widely known as the creator of "defensible space" principles, considered the gold standard for how communities can protect homes where wildland and human development meet—the wildland-urban interface. Findings from his pioneering research call out an overreliance on wildfire suppression, as well as a "false narrative that community protection requires wildfire control."[13] Moreover, he and his colleagues have unearthed dimensions of fire behavior undetectable by earlier models.

Working with other post-wildland fire investigators, Cohen studied both drainages and developments to reconstruct fire behaviors from charred scenes. He looked for clues revealing how a wildfire sped up or slowed down in certain spots, why it chose to incinerate certain fuels and left others intact. He also studied human endurance and reaction times in relation to moving wildfires. Following the 1994 South Canyon Fire that took the lives of fourteen firefighters, he ran up slopes where firefighters had attempted to get over a ridge to safety, assessing their reaction time as they were chased by the inferno.[14]

His mode of thinking was inductive, not deductive "mathematical modeling in the woods." He drew inferences from the details he observed and pieced together a larger picture. In this approach, he could compare what fires "wanted to do" with the quirks of human judgment in complex conditions. Vegetation, topography, weather, and witness accounts gave clues to dissect the gap between what fallen firefighters thought they were seeing (and were reacting to) versus what the fire actually did. Their sense of the speed of the oncoming flame could also differ from its actual speed. But it wasn't up to them to overcome sensory miscues with better fast thinking.[15] And there's the rub. The onus was on decision makers who had to scale back their confidence in wildfire suppression and account for an unavoidable disconnect as a core issue of firefighter safety.

But recommendations derived from inductive research were not always welcome. After the 2000 Cerro Grande Fire in New Mexico, Cohen went house to house assessing the destruction of homes and other structures. A coworker who accompanied him described how "Jack was just watching," seeing how houses burned, why they burned, and the apparently random nature in which one house burned and another one did not (in his words, "one of them burns over here, all by itself; a neighborhood would burn over there and over here, nothing would happen"). Cohen concluded that, in this case, it wasn't the wildfire that burned down the houses. It was, in fact, the tiny embers and creeping ground fires igniting receptive fuels (the litter accumulating on chimneys, roofs, gutters, and decks). Those embers also circulated in interior vents and crawl spaces, making homes ignite from the inside out. In a report he filed about the Cerro Grande Fire, Cohen wrote that "the high ignitability of most of the residential area allowed numerous simultaneous house fires that quickly overwhelmed the suppression forces."[16]

For an agency focused on suppression (increasingly, in the wildland-urban interface), "overwhelmed" was not a good word. The observed combustion patterns were at cross-purposes with institutional framings of suppressable fire. Yet the problem of how communities can coexist with ecologically appropriate fires remained. Cohen derived critical interventions from this core concern, among them the well-known "home ignition zone" concept, which is based on the idea that the owners of homes situated in fire-dependent ecologies had a role to play in reducing wildfire risk. If receptive fuels were removed, if structures were treated more like blocks of wood (which take longer to ignite), they would more likely be spared. Maintaining physical compatibility with the surrounding environment was "just mostly housekeeping," as the coworker put it to me. Housekeeping of this sort would also obviate the need for more suppression as the automatic response, which can add unnecessary risk. The Forest Service's mission statement ("to sustain the health, diversity, and productivity of the nation's forests and grasslands to meet the needs of present and future generations")[17] does not mention protecting homeowners' private property, "but these things are conflated all the time."

The Circular Nonsolution Space

Cohen's research showed how knowledge production around wildfire filters through vested interests that shape certain visions of fire so that, in his blunt assessment, "We don't have an intellectual perspective on something that is kind of a big deal—fire." Calling out this lack of perspective, he remarked: "If you can go to the moon, which is just a complicated engineering problem, then you can do anything, except understand the planet that you live on or accept how the planet that you live on works." Certain economic mantras keep fueling the vacuum of "not knowing what's going on with regard to fire," inviting spurious certainties, "like little economic statements, such as the smallest fire is the cheapest fire," that often fall short of reality.[18] His colleague Finney would attribute this state of affairs to a wildfire science that, at least in the United States, became "one of those areas in which the self-correcting part of science doesn't really get fulfilled." For Cohen, one additional factor undermines this enterprise, which is that in continuing to justify convenient images and mantras of wildfire control, we have lost an "ordinary ability" to keep ourselves from "being imposed upon by gross contradictions."

Under these less-than-ideal conditions, Cohen was "wait-ing for a reality response, rather than a fatalistic one; an intel-lectual response, rather than a visceral one." The fatalistic and visceral ones reflect a view of fire as something always to fear or eliminate, making some people "quit thinking." When children reach for a flame on a stove, they learn to control fire risk by withdrawing their hand from the flame.[19] Yet the supposed mastery in exercises like these (in which "we're supposed to be excluding fire and not supposed to be playing with fire") becomes counterproductive in wildfire science; it reinforces "a monument" of cognitive dissonance.

When I asked him to elaborate further on the dissonance, Cohen pointed straightaway to his hand. He mentioned that a flame exposure that would generate a second-degree burn on a human hand "won't even char a block of wood." In other words, human senses operate in a totally different range of sen-sitivity to pain and injury than "what it takes to ignite some-thing" in wider combustible worlds. Given this difference, they can generate their own sort of miscues, turning sensate bodies into dubious instruments for apprehending risk. This embod-ied baggage does not just fade or disappear; it shapes a certain "experiential relationship" with fire that follows firefighters into the field. As extreme wildfire behavior becomes more frequent, more firefighters will be recruited. His concern is that for those in the field, firefighters will need more, not less, support in the form of lookouts and observers who can spot and monitor wildfires, making sure that everyone is safe—that is, that every-one is seeing the same fire.

Remediating knowledge (and sensorial) vacuums requires more than a shift in scientific approach (from deduction to induction, for example). It entails the recognition that illusions of control have consequences for the future. As long as fire can give rise to different (subjective, visceral, fatalistic) expectations and perceptions, these can be stirred in ways that allow vested interests to promote certain features about wildfire controllability while downplaying others.[20] In calling out the lack of intellectual perspective on something that is "kind of a big deal," Cohen is pointing to a level of operational precarity that will need constant shoring up, particularly as disasters mount.

It is important to note that the antidote to suppression in its US form is not antisuppression. Among other things, it is active curiosity as a value proposition that allows institutions to change methods and protocols to get on a different trajectory, enabling fire science to support the reality of those changing conditions, rather than deploying anachronous slides and more resources (i.e., people). Former wildland firefighter Timothy Ingalsbee suggests that, "instead of reactively cutting fire lines in a suppression state of emergency," firefighter roles and identities should shift to engage a "full spectrum of possible roles and duties" in ecological fire management.[21] At the same time, Indigenous fire sovereignty should expand, allowing Indigenous practitioners to engage in proactive biophysical stewardships while setting standards for non-Indigenous fire management and offering an opening for stemming the damage. Despite legal prohibitions and barriers against cultural burning, these observers of the non-Indigenous observers of fire are attempting to live with fire in increasingly hostile climate futures (see chapter 6). In the aftermath of the Castle Fire that consumed a mountain ridgetop in California in 2020, Bill Tripp, director of natural resources and environmental policy for the Karuk Tribe, noted that it "cost $100 million to fight the Castle fire. What if we dedicated that $ to restoring our forests to reduce fires in the first place?"[22]

Feeding unrealistic assumptions about wildfires and how they can be "contained"—presuming, against evidence to the contrary, that tomorrow will look like today—will only cost more lives down the road. It cuts off what I have been calling horizon work, in which a set of diversely positioned constituents can partner to forestall what Cohen calls a "massive ecological mistake." That mistake isn't just about a dire climate emergency; it is an accretion of ecological impacts from misguided policies that, enacted under extreme conditions such as those that exist today, create catastrophic conditions that increasingly refuse solutions of any kind.

Little Green Pine Needles

In the meantime, what can a commitment to scientific inquiry amid "monumental" cognitive dissonance look like? Earlier I mentioned that the strong personalities that set the direction of wildfire science in the United States were steeped in, and at times clung dogmatically to, mathematical deduction—even when outcomes of that deduction could be inconsistent with physical reality. The inconsistency was tolerated, even for something as basic and consequential as the mechanism of wildfire spread.

To see this mechanism, imagine a big, flat wall: when enough energy is transferred to it, it will ignite. The heat transfer mechanism in this instance is called radiation, in which ignition occurs without direct contact with flames (the air transfers heat from flames or burning objects to adjacent fuels). In structural fire protection, radiation explains the combustibility of objects (like big flat walls or furniture) that are compact or solid and have continuous surfaces. But wildfire typically spreads in discontinuous and variable surfaces—in fine or loosely packed vegetation, such as pine needles, leaves, twigs, grasses, and bark. Though radiation was an assumed heat transfer mechanism here, it was never sufficiently validated. And if radiation doesn't always explain how wildfire spreads, then what does?

Early on in his career, Cohen (and later Finney and other colleagues) saw that physically meaningful descriptions of wildland fire heat transfer—and the experimental studies from which those descriptions could be derived—were missing. Conventional fire-spread modeling assumed a governing heat transfer mechanism that also governed the kinds of questions that these researchers could ask. Alternatives, like convection, a form of heat transfer in liquids and gases, had been explored, but not enough. When Cohen raised the possibility of convection with a dominant wildland fire modeler at the lab in 1981, the modeler shut it down right away, saying that convection can't be responsible for wildfire spread and, "if that was true, it'd all burn to a shockwave!"

By that time, however, Cohen had already left the Missoula Fire Lab and moved to Southern California, where he worked at the US Forest Service's Riverside Fire Lab. Committing himself to studying phenomena that had become "physically inconsistent with perception," he first focused on so-called live fuels (or living vegetation). Live fuels were of particular interest because they don't burn the same way dead fuels do. He knew he had to do a lot of observing to be able to come up with effective questions about their ignition properties. So he farmed himself out as a fire-behavior analyst to local districts so that he could run prescribed burns. He took on a job as a lightning supervisor for the San Bernardino National Forest, and he studied fuel breaks—barriers to wildfire spread made of damaged soils or bulldozed ruts where nothing supposedly burns, including live fuels.

He focused on chamise chaparral, a common shrub, which grows in those fuel breaks. When he set a bit of it alight, he watched the supposedly inconsequential little shrubs burn.

Fire-spread models skipped over the fact that the needle-like bundles that sprout from them "will carry fire all by themselves quite nicely."[23] That is, the fuel breaks weren't working.

Cohen reported his findings on fuel breaks and how they work (or don't work) to the USFS Washington office in 1985. He questioned whether the models assumed unwarranted confidence in fuel breaks. The report, he said, got shelved, and he was given a "directed reassignment" to Macon, Georgia.[24] But by then he had dismantled some hypothetical universes: where models had discounted live fuels, they in fact burned, albeit in different ways. This insight was important: after all, how was it that live fuels, like little green pine needles, could become the source of explosive crown fires that jump across and incinerate the living canopies of trees and nearly everything else in their path? Little green pine needles, and how they ignite, were key to unlocking nonlinear, abrupt, or explosive fire behaviors.

Cohen returned to the Missoula Fire Lab in 1996 and began collaborating with Mark Finney, who had arrived there in 1993. Physical inconsistencies with models were now harder to tolerate. Massive crown fires were becoming major threats in the field. But no one could ever get little green pine needles to ignite in a lab. A crucial question remained unresolved: What makes for a sustained ignition for these so-called fine particle fuels, the source of the most intense (and inoperable) kind of wildfire that incinerates conifer canopies?

They did the obvious, which was to focus high radiant heat on a single pine needle, trying to get it to ignite. Under the same temperature, within twenty-seven seconds, a small wooden block would burn. Within thirty-five to forty seconds, a wooden wall. They kept the heat on it, but still the pine needle would not burn. With the help of a Missoula-based sculptor, they built an apparatus that exposed the single fine particle to now-scorching heat

FIGURE 8.1. Remains of fire-damaged trees stand in the aftermath of the 2020 Beachie Creek Fire near Detroit, Oregon (Reuters/Shannon Stapleton).

(up to eighteen hundred degrees Fahrenheit). An ultrathin sensor (thermocouple), placed on the pine needle's surface, measured temperature fluctuations at a rate of five hundred times per second. At such high resolution, what the particle was doing, right up to and at the point of ignition, could now be observed: the surface of the particle was heating up and cooling down in pulses. And when it finally ignited, it did so "all within two seconds."

What does all this have to do with failing fuel breaks, the treetop-to-treetop jumps of a crown fire, and wildfire behaviors more generally? Cohen's onetime colleague, Jason Forthofer, an active wildland firefighter and researcher, told me that firefighters routinely see phenomena in the field that the models don't see. Among them are pulses: flames that overshoot, accelerating past a point of some expected spread rate, and then slow down. Making sense of them would enable Forthofer to account for rather than discount them to better predict what a fire might decide to do next.

Subsequent fine-fuel experiments continued to explore a messier mechanism of wildfire spread, past rate of spread questions, and beyond the notion that humans can control nature, or tip it back. One experiment led to a 2015 landmark paper that showed how convection, not radiation, is a key mechanism for wildfire spread.[25] On one visit to the lab, the researchers had set up a small version of the experiment so I could observe the convective forces at play. A sloped test bed contained laser-cut cardboard pieces, arranged like combs to form a table of "fuel." When I lit the fuel at the lower end of the table, flames initially rose and then steadily moved up the test bed.

A few seconds into the burn, Forthofer asked me to observe the overshooting dynamic that had just kicked in. The fire was jumping. The peak-and-trough formations of the moving fire became more pronounced. Look at one of the bigger troughs, he said; there seemed to be some force pushing down on the flame, causing a rotation that kept the fire moving horizontally (and along the test bed). A still from a time-lapse video of the experiment shows vortices that are indeed rotating within a trough (see fig. 8.2). They push the flame upward, downward, and then forward in a subtle but noticeable pattern.

All the while, the flame needs air to keep up this pattern. It will jump ahead of itself to take in air. If that air isn't available, it will suction it from its sides. And if that air isn't available (as in, say, a real-world drainage-type situation, where less side air is available), the fire will pitch forward and race through the drainage in a frenetic attempt to secure air. In 2017, Forthofer, a seasoned professional, had skirted one of those instances when "any little spot fire could start in the drainage—there's steep walls on

FIGURE 8.2. Still from fire experiment showing forward bursting and vertical patterns of flame peaks and troughs (Finney et al. 2015).

both sides—any stalling of time, any being oblivious for another five minutes would have been a costly mistake." That time, the "fire just didn't happen to react."

Forthofer asked that I look at the flame again—how is it spreading? Another insight in this round of observation, also important for anyone in the field, awaited. What was happening inside the troughs was, to some extent, happening *independently* of the fuel bed. This independent activity suggests that in some cases there will never be a realistic basis for wildfire control: suppression forces will be overwhelmed. As the influence of climate change on fire weather conditions becomes more pronounced, pulses and other erratic wildfire behaviors have greater latitude to evolve. They can also reach up into fire plumes, combine with the atmosphere, and produce infernos on much greater scales. While small-scale experiments are a long way down from these atmospheric processes, they tell us about how the physics of wildfire can couple with climate chaos, perpetuating massive ecological mistakes that risk becoming the only physics that we see.

9

Going through the Porthole

Unlike some of their counterparts in other countries, US wildland firefighters work in close proximity to wildfire. They are trained to remain aware at all times of escape routes, nearby clearings, and safe spots that have already burned over (known as "the black"). To decrease chances of being overtaken by unexpected fire behavior (known as entrapment), they are trained to "keep one foot in the black"—or to maintain access to a designated safety zone.[1] They deduce the size and location of a safety zone from flame geometries and on the basis of continually updated guidelines for how to make a deduction.

The utility of such guidelines can be compromised by new circumstances. As these are more readily felt, safety practices and enhanced situated awareness become less reliable. Sometimes, there is no black, or, with unburned fuels, drought conditions, and rising temperatures, even the black somehow burns. Crew members can lose their way in the shifting smoke and flame, and so they routinely depend on others (safety officers, squad leaders, lookouts, superintendents, and so on) to radio them about what is going on or what to expect next. When information is limited or if communication breaks down, they must focus on getting to a safety zone or, in a worst-case scenario, prepare for imminent wildfire entrapment.[2]

In what follows, I take the reader into aspects of this unwanted scenario, where a messy experiential relationship with fire meets the challenge of self-preservation. I center my attention on the risk of entrapment: how it can be prevented and how survival chances can be enhanced. While suppression operations draw in thousands of firefighters annually, wildfire entrapments resulting in fatalities are rare (aircraft and vehicle accidents claim the most lives). But as the frequency and distribution of large wildfires increases and more people are involved in the battle against them, the issue of entrapment becomes a more salient concern. Building on insights from people who have either experienced or investigated entrapment, this chapter privileges their perspective to explore how life-threatening dangers and their potential warning signs are both felt and rationalized.

In the United States, facing fire entrapment means resorting to an emergency fire shelter, which the Forest Service began developing in the late 1950s. By the mid-1970s, the agency had made carrying the device mandatory for its firefighters.[3] Made of a layer of aluminum foil backed by a heat-resistant silica cloth on the outside and, on the inside, aluminum foil laminated to fiberglass, the shelter reflects radiant heat and slows heat transfer to its interior. Amid hot and asphyxiating gases, it also gives firefighters a precious volume of breathable air. As I set out to learn more about the scientific and technical challenges of firefighting, fire-behavior analyst Jason Forthofer referred me to his colleague Tony Petrilli. A former smokejumper and equipment specialist who works at the Forest Service's Missoula

FIGURE 9.1. Inside a safety zone (https://wildfiretoday.com
/tag/safety-zone/).

Technology and Development Center, Petrilli believes that survival instinct, and finding ways to instill it, eclipses other ways of objectifying risk and keeping firefighters safe.

Survival can come down to a "coin flip," especially when the onrush of heat, wind, falling embers, and choking smoke makes it difficult to determine if conditions are safe to deploy a shelter. In 2013, a fast-moving wildfire approached Yarnell, Arizona, killing nineteen highly skilled firefighters who had deployed fire shelters. They were in a box canyon that blocked their view when they were overwhelmed by a fifteen-mile-per-hour wind that pushed flames parallel to the ground "in a ton of direct-flame contact." Their deployment sites were not survivable. A postaccident investigation of the Yarnell Hill Fire noted surrounding temperatures of over two thousand degrees Fahrenheit.

Petrilli himself survived a disastrous blaze (the 1994 South Canyon Fire) by deploying a fire shelter. In his words, he "went through the porthole" and survived; fourteen of his fellow firefighters did not.[4] My conversations with Petrilli made me appreciate how the risks of firefighting are not always easy to objectify. It can be challenging for crew members to even see the same thing—be it in the fire (is the wildfire moving faster or slower?), on the ground (has the fire burned out here?), or in the distance (how close is the fire?). Such individual differences in perception, while not unusual, can certainly undermine the shared understanding on which coordinated fire management crucially depends.

Talking with him placed the variable aspects of fire that I had been learning about in the Fire Lab in a context in which little, if anything, lines up with expectation: "Weather is a big factor in fire behavior," Petrilli said. "Temperature doesn't change very quickly, but wind sure can. Wind speed can increase, or wind can change direction—and that's probably one of the big ones—and underestimation of fire behavior, I mean, that [element] is in all of them." Fire does not spread according to averages—or, as another specialist put it, "It's not that the fire is coming this way and it's moving at one mile per hour and it's got a good wind on it.' It's more like 'If it changes direction, it could come at you at five or ten miles an hour and you have no chance of getting out of the way.'"

Eric Hipke, a smokejumper who, like Petrilli, survived the South Canyon Fire, described the feeling of being struck down by something like a wave: "The heat was so bad, and I don't remember exactly . . . it's a little fuzzy . . . it's like I tripped. I was jumping to get away from the heat and it felt like . . . like a wave hit me. It was like . . . being at the beach and not paying attention, you know, standing with your back to the ocean and a wave comes and just knocks you down; so, I ended up on the ground . . . Investigators told me they thought that if that air that had . . . knocked me down had been superheated—that's what saved my lungs from getting fried. I immediately got back up and my hard hat was off, got knocked off, my pack—either the straps had melted off or the straps had slipped off . . . I was waddling up, I just punched my pack loose, and I just ran over to the top of the ridge."[5]

Petrilli's number one concern is safety. He has been fighting fires with the Forest Service since 1982 and has been on more than thirty investigations of fire entrapments. He is attuned to modes of rationalization that can heighten risk in the field. If, as in the prior chapter, fire researchers are trying to understand fire objectively, a reasonable assessment of firefighter surroundings is sometimes impossible no matter how much knowledge he or she brings to an actual wildfire. "Everyone sees fire differently," he told me, "and they will react differently" based on what they see.

As he reconstructs events leading to entrapment, he tries to imagine firefighters' thought processes as they decide to retreat or deploy the shelter. Some might drag out the decision, believing that entrapment "can't happen to them." Petrilli calls them "last-fact thinkers." After seeing enough fatality sites, he wants to see more firefighters who are "first-hint thinkers," acting on the first hint of danger, not waiting for the last fact.

To illustrate the difference, Petrilli told me about an entrapment in California that had happened a few weeks prior to our conversation. Some of the crew members wanted to leave. Other people said " 'no, let me go take a look at it to make sure it's what I think it is.' They're not being superheroes," he said. Nor are they at fault for their decision; they are just thinking, "it will not happen to me."

In the calculus that Petrilli would like to see informing firefighter decisions as to whether to press on or retreat, first hints should function more like objective facts, and last facts should be treated as just wishful thinking. Switching to a different scenario, he described two firefighters observing an oncoming flame. When they sense they will be overcome, they look to the ground beneath their feet. One of them thinks it is sufficiently burned over to warrant deploying a fire shelter. The other says that the deployment isn't going to work: "too much brush, too much fuel, too many trees in the area. We've got to go."

The first firefighter stays and deploys a fire shelter; he is hoping that in his coin-flip he will still have a chance to survive if the ground does not burn. Once he realizes he was wrong, it is too late. The second one runs down the mountain to a road, gets picked up by his crew, and survives. When the accident-review team finds the remains of firefighter 1, they see that the fire, in fact, had crept in from beneath the shelter. The soil wasn't black enough—that is, it could still burn. Petrilli insists that both would have been right because they arrived at decisions that made sense to them at the time.

There was only so much certainty to be had. What Petrilli cares about most in this circumstance is the firefighter's ability to stay alert. First-hint thinking offsets firefighters' overthinking a situation (or last-fact thinking). This greater risk might leave them stuck in precious time-consuming questioning or with a shelter left undeployed. Firefighters are always negotiating potential or unseen hazards, but last-fact thinking is like quicksand, a dangerous point on a trajectory toward mentally shutting down. That is why Petrilli cofounded a program, "Fire Shelter Deployments: Stories and Common Insights," in which firefighters share entrapment stories to help others stay alert to "what they may experience . . . and how to survive."

Petrilli reminds firefighters that their lives are not expendable. There is a crucial distinction between the risk they add to their lives and *risking* their lives. He knows it could be easy for them to bow to the pressure of what a threatened community might have them do—accept more risk and possibly risk their lives. Petrilli uses different strategies to inhibit firefighters from stepping into this quicksand. He presents recruits with ordinary thought experiments, such as asking them whether they would accept more risk to "contain a little fire in a green pocket of unburned area [far] from the fire's edge." Even if it burns, he tells them, the little fire is not going to escape. "But there are some people who still want to go in there and mop it up because there's smoke, or the people living down in the valley are going to call and say, 'Hey, there's still smoke up there.'" They did not sign up for a suicide mission.

Such is the cornerstone of the fire shelter training that firefighters receive. Yet sometimes they are resistant to deploying a fire shelter even though, under the right conditions, it can save lives. Why? Because deploying a shelter might be read as a failure. "It's harder," Petrilli says, "to do nothing than to do something." To help shake their resistance, Petrilli presents other thought experiments. He asks firefighters to imagine a worst case, in which an accident-investigation team finds their hypothetical fatality site. Once the site is found, an investigator will look for clues to reconstruct a crew member's thought process leading up to the fire entrapment. At the imaginary fatality site, he reminds trainees that "you want to show that your decisions made sense." He asks them, "How do you want investigators to characterize the site? . . . If we accept [a certain level of] risk, how is the accident report going to read? We were up there doing *what?* I use that one a lot."

Mike Cooper, also a smokejumper at the South Canyon Fire, managed to survive wildfire entrapment by deploying his shelter. In a safety training video, Cooper describes the challenges of doing so. To reach a solid surface on which he could deploy the shelter, he had to scrape away six to eight inches of leaves, soil, and dust. Amid the high winds, "every leaf, the dust, the soot—anything loose—was actually getting sucked into the fire, like a giant vacuum cleaner." When it came to opening the shelter, "you didn't just flop it open like a nice loose blanket, shake it a couple times like in the training video, and crawl into the thing." Cooper continued, "[The shelter] is streaming out from you like a flag whipping in the wind as you're trying to open this thing. It's just flapping crazy . . . , one loose grip and you would've lost it."[6]

In these moments of desperate fumbling and uncertainty, he recalled in the training video how easily firefighters can "mentally shut down." Between the "noise and the confusion and the panic . . . you realize you're going to crawl into this thing as a last resort . . . It's almost like crawling into your coffin to see if it fits before you get your burial. It's a pretty strange psychological feeling when you realize you're committing your last hope to this piece of tinfoil."

Petrilli knows the shelter and its limitations like nobody else. After the South Canyon Fire, he and his colleagues redesigned its handles to make it readily deployable through a single-motion release. Since his entrapment experience, training firefighters in the use of shelters remains his life's mission. In assessing numerous fire entrapments, he has seen the "million ways" of saying "I don't need [to deploy the shelter] yet" and the consequences of resistance to its use.

In one of our conversations, he showed me a video used in wildfire-safety training courses; it demonstrates how instances of "waiting for facts to emerge"—when firefighters think objectivity "is just down the road if they wait a little longer"—can manifest. Produced by the "Fire Shelter Deployments: Stories and Common Insights" program, the footage came from a 2014 wildfire in drought-stricken northern California. Firefighters were operating bulldozers ("dozing out a safety zone") when the fire switched directions and headed their way.

In the video, one of the crew members films the approaching fire with his cellphone. Others make comments about the fire while holding fire shelters in their arms, or take photographs. A contracted bulldozer operator is moving combustible material to create a safety zone. Burning embers are starting to fly everywhere. The firefighters will survive the incident, but they are waiting too long to deploy their shelters.

A few minutes into the video, the flame front crosses a nearby trail and starts coming at the crew. Soon enough, dust, dirt, and ash start to swirl around as an approaching wall of flame begins to suck in the surrounding oxygen. The crucial moment to deploy has arrived. As we watch the clip, Petrilli comments, speaking with a sense of urgency and care, "I mean, it's coming. I'm ready for them to get in their shelter, right now." The waiting continues: some are still filming while others are working on the safety zone. They are thinking, "we're okay, we're okay, we're okay." The bulldozer still hums in the background. The dozer operator might be thinking he is following protocol—in this case, by digging out a bigger safety zone—and that it is going to help. As this sobering scenario continues to unfold, it becomes evident that the firefighters will soon be entrapped. Yet right up to that point, they are pressing on—or they are in denial, or underestimating the speed of the fire, or relying too heavily on mental slides from previous fires, or all of the above.

The Ten Standard Firefighting Orders (SFO) is a decades-old set of guidelines that reminds firefighters of situations that require constant monitoring (fire weather conditions, fire behavior, identified escape routes, proper communication, and "fighting fire aggressively, having provided for safety first"). Introduced by a Forest Service task force in 1957, they are also the stuff of an early warning system that can permit timely knowledge of dangerous and safe conditions. Rule number six reads, "Be alert. Keep calm. Think clearly. Act decisively." In his witness statement for the South Canyon Fire's investigation report, Petrilli attested to following all the rules except one: "We did [all of] these except stay alert."

Today, the wildfire community benefits from the "wildfire lesson learned" videos he has produced; those who agree to be interviewed about their entrapment experiences do so with courage and a resolve to help others reflect on the stigma of shelter deployment and how illusions of safety and wishful thinking can make firefighters vulnerable. Indeed, such stories

Video: HEQB (t)   Entrapment site   Time: 1734-1737

FIGURE 9.2. Fire entrapment, training video still ("Fire Shelter Deployments: Stories and Common Insights," https://www.youtube.com/watch?v=z7EwGSZQo0I).

and images provide powerful object lessons, reminding new recruits that firefighters overtaken by fire are just like "anybody and everybody," in Petrilli's words, and that "it can be you" who thinks it isn't going to be all that bad. Such lessons turn a problematic experiential relationship with fire into a set of stories that can counter hesitancy and delay in recognizing imminent entrapment.

When Petrilli came out of his fire shelter in 1994, he looked around to see who else was alive. As one journalist described the moment, "His elation at emerging a survivor didn't last long. Within minutes, he was among the first to find the bodies of some of the 14 firefighters whose fire shelters didn't save them. His radio message reporting the deaths rattled federal agencies," leading, among other things, to a 2002 redesign of the fire shelter in which Petrilli played a pivotal role.[7]

The deadly tragedy of the Yarnell Hill Fire prompted the Forest Service to expedite the fire shelter's next redesign. The goal of this redesign was to improve shelters' "ability to withstand direct contact with flames."[8] In this redesign effort, the Forest Service partnered with engineers at NASA's Space Technology Program who were developing heat-resistant shields to protect spacecraft as they decelerate from hypersonic speed and reenter Earth's atmosphere. Even these thermally enhanced materials do not withstand some of the extreme fire temperatures encountered on Earth.[9]

The search for improvement continues, but for now the first (2002) shelter redesign is hard to beat in terms of protection and weight. We were in a sprawling equipment-fabrication shop, talking over a table where an off-duty smokejumper was cutting pieces of a promising fabric obtained from the Latvian space program. Petrilli showed me a video from an experiment at a University of Alberta outdoor lab, meant to test new fire shelters under a simulated high-intensity crown fire. Eight propane torches baked the aluminum-encased shelters with direct flame contact; the heat from the torches was still not nearly as great as that of a crown fire.

It was hard to believe that anyone could survive that. The old shelter design lasted only about ten seconds in direct flames. The 2002 shelter offers about a minute of survivability as it slows the heat transfer in up to a two-thousand-degree flame. He hoped to add another thirty seconds with the promising fabric, which would be shipped to the Canadian lab. Ultimately, the new fabric experiments did not yield the increment of improvement Petrilli and his colleagues were looking for.

Firefighters in many other countries do not get this close to fire, nor do they necessarily want to. A representative of the Australasian Fire and Emergency Service Authorities Council stated, "Our experience is that you need a solid barrier between heat and a person." Fire officials in Australia have been quoted as saying that the very availability of shelters can induce

FIGURE 9.3. New-generation fire shelter test in Alberta, Canada: high temperatures melted the aluminum covering, exposing heat-resistant silica cloth (2015) (photo courtesy of Ian Grob/U.S. Forest Service).

"firefighters to think that an item like that will protect them."[10] In the state of New South Wales, tankers double as shelters and have a spray system to extend survivable conditions.[11] In France, in the case of an overwhelming flame front, vehicles serve as last-ditch fire shelters. French firefighters are not issued fire shelters; instead, they carry "an emergency smoke mask for retreating back to the vehicle," implying that they will stay close to a truck and drive away rather than deploy a shelter.[12] In certain provinces of Canada, "firefighters are never put in a situation where they would need to deploy a fire shelter."[13]

Petrilli is cognizant of these national differences. Because wildland firefighting in the United States can entail direct confrontation with fire, the shelters, he and his colleagues believe, are necessary. Surveys indicate that firefighters want to carry

FIGURE 9.4. A firefighter wears a heat-resistant uniform in Rafina, Greece (Costas Baltas/Reuters/2018).

them. The use of the fire shelter is spreading globally, particularly to countries whose national fire agencies are seeing more fatal entrapments in contexts of extreme drought and missing escape routes and safety zones.[14] The shelter does not eliminate the need for other equipment for surviving the fight with wildfire, such as self-contained breathing apparatuses and better means of countering dehydration, heat-stress-related injuries, and other health risks in the field. Despite the fire shelter's limitations and ongoing debates as to whether it adds protection or induces a false sense of security, shelter deployment training and messaging with respect to its proper use continues. In the meantime, stories of fire shelter deployments continue to be publicized, reminding firefighters that this one last chance for survival is just that—a chance—not a portable safety zone, or a Superman suit.

10

Beneath the Airshow

Dismantling the image of Superman requires more than just operational tweaks. In the wake of the deadly 2013 Yarnell Hill Fire, one firefighter noted how tweaks by definition "have almost no perceptible impact because they nibble around the edges of symptoms." In an online essay that circulated widely in the wildfire community, wildland firefighter Mark Smith addressed his words to fellow "hose-draggers, fire directors, dirt diggers, academics," and agency administrators. He reflected on what he called the "big lie," defined as the denial of underacknowledged safety issues in wildfire suppression, and he sought to make sense of firefighter entrapments and to honor the dead.[1]

Underscoring changing conditions on the ground as threatening firefighter welfare, Smith pointed to a "disconnect between reality and action." He quoted one of his fellow firefighters, who said he "can't help but feel that there is a [needed] conversation . . . about our mission as suppression resources. Are we now in the business of intentionally risking lives to achieve wildland fire objectives? I ask because at least the [agency] has never accepted that position before and maintains its stance on zero tolerance to this day. I understand that firefighters are going to die but there is a big difference between vehicle accidents and entrapments." The calculus shifted toward risking lives, but that truth is being "stifled by the denial [that's] happening right now." The essay struck a nerve because it raised a question to leaders about what exactly they are training firefighters to do. Should crews be trained as resources chasing suppression fires, and what justifies putting people in dangerous positions in the first place?

For the many professionals teaching tactical skills to new firefighters, the search for justification is ever more pressing. Some spent years working up and down chains of command, and were acutely aware of how an increase in unsafe conditions was leaving firefighters exposed. Their concerns centered on firefighter safety zones and whether these could still be secured. For former crew leaders, these were places of refuge where a firefighter could watch a wildfire from afar. Active leaders also held on to the idea of safety zones as "safe," even though they were clinging to a concept that at times offered only an illusion of protection. This gave them all the more reason to stand firm on the principle of safety. As one leader told me, when safety zones cannot be secured, no suppression resources, including firefighters, should be deployed. There should be no ambiguity: a safety zone is a place you can go and hide—"there is no concern—zero."

No one could disagree with this view. But how that nonnegotiable line would be drawn was the issue. Extensive training and guidelines aim to maximize the safety of firefighters. Yet many see their duties as being far afield from those implied by their job title, which is actually "forestry technician."[2] It's an odd title, harking back to earlier times when, as one forty-year-old firefighter told me, "we did lots of other work, like trail clearing and road clearing and thinning and non-fire-related forestry work." But one firefighter embraced the seemingly outmoded designation, telling me, "We are forestry technicians. We work in the forest to manage a sustainable agriculture crop, which is trees. We're here to do what's best for that crop and what's best for the forest." Today, the nominal ecological protector works in conditions that are increasingly more difficult to (ground) pound into a stable state.

From his vantage point, this responder observed the workloads of crew members from the Southwest with particular awe. He wondered when he would have to deal with similar workloads. For example, in 2000, a year of devastating wildfires in the western United States, he said, referring to these crew members, "We saw some of them outside of our town in early summer, that was early in the fire season. And they were already at twelve hundred hours of overtime versus ours—I think at that point we were at two hundred hours." The Southwest crews "just kept going—their whole world became nothing but 'go do this.' It's like robots out there by the end of the summer."

Some firefighters recalled a time when they were not robots, and when patterns afforded a predictable sequence of demands. An incident manager painted a similar picture: the monsoons would come in the Southwest and then the fire season was complete; after that, "the Southwest firefighters were freed up to come up to the Northern Rockies to help when the fire season started there a bit later, or to the Northwest or to California." Unpredictable fire seasons have led to exhaustion, so much so that bipartisan senators introduced legislation that would rename "forestry technicians" as "wildland firefighters," giving them job stability and access to benefits. The legislation would also remove the word "seasonal" from their job title, allowing them to work beyond the previous limit of 1,040 hours per year.[3]

The so-called pack test that measures a wildland firefighter's ability to perform strenuous work for a given amount of time would also need adjustment. At the time of my research, these fitness tests were being adapted to better align with the physical demands of firefighting.[4] The physiological effects of overstress became acute, as one recruiter told me, pointing out the toll inflicted when "you work your body to a point where it can't recover." At that point, "the body will consume itself; once the fat's gone, it eats the bone." She was describing a serious but underreported condition called rhabdomyolysis, in which an exposure to high heat can trigger a surge of creatinine production that, in turn, leads to muscle tissue breakdown and heart and kidney damage. "It looks like heat exhaustion," she said of a disorder that is on the rise. With the hope of improving scientific understanding of the problem, some have volunteered as test subjects for physiological studies of prolonged exertion and strain in harsh conditions. Underdiagnosed "rhabdo," as it is called, points to conditions in which firefighters exert themselves to burnout and beyond.[5]

The loss of life weighs heavily on the wildland firefighting community. One midcareer firefighter tested my recall of a particular fire season. "Well, there was the Idaho fire, and then the massive one in California, and Fort McMurray, Alberta," I said.

"The big one," he seconded.

"Yeah. But that's about it," I said.

"Pretty much all you saw on the news, right?"

"That's most of what I heard on the news."

The news cycle had missed a wildfire incident that had claimed the life of a young hotshot in Nevada. "So, stuff happens," he said. "But it's not like a giant fire season, like the year before."

Owing mostly to high winds, the Strawberry Fire grew from fifteen to roughly forty-six hundred acres in just forty-eight hours. Called in three days after an initial lightning strike, the cause of the fire, a rookie sawyer tried to undo a "tangled tripod of trees" to help stop the fire spread. In the process, a twenty-six-year old from Vermont was struck by a snag tree that was "leaning into the forked top of the support tree and was burning at the stump," according to a fire fatality report. After he applied a series of cuts, "the top of the snag pivoted violently out of the fork of the support tree." Approximately fifteen feet from his cutting position, it struck the sawyer, killing him with blunt-force head injuries.[6]

In the report from which I have quoted, there is a section on "sensemaking" that asks readers to consider "what seems important to attend to," given the details of the accident. One of the report's subsections addresses a course of events in which a "drive to work faster, quicker, or easier can take over." Its title takes the form of a question, "Can We Say No to a Tree?" When sawyers approach a tree "to size it up, they have inherited all the higher-level decisions" that have brought them to a position of "assessing a hazard tree for removal." The sawyer "alone was closest to the task." He is depicted as capable of making his own assessment because he had "the best information to make the decision" about completing the cuts. The sensemaking section does not pose the question of why the sawyer was there, on his own and without an extra set of observers, in the first place. And that is a question worth asking. A higher-level burden of decision making was shifted onto him, which returns us to the question about who draws the line between safety and risk. Could he have made some calculation to give himself permission to leave? Did he really have the best information because he was closest to the task?

FIGURE 10.1. Air tanker drops flame retardant on the 2015 Valley Fire in California (Shutterstock/Kirstin Adams-Bimson).

The Airshow

In the year of the sawyer's passing, concerns around the mission of suppression resources surfaced again. When officials who command the resources to suppress wildfires decide to "throw everything they have" at them, as is often said, it usually means an "airshow" of flame-retardant-throwing air tankers and every instrument of suppression available. What happens just beneath them?

The question returns us to the crew leader who stands before a wildfire and wonders whether she should send her crew to fight the fire. She is not like the commander who leads a crew into a burning house, only to save them with fast thinking. Our commander doesn't have that kind of power. Someone has dispatched her crew to be there. But she doubts members' ability to control whatever the fire, fueled by drier-than-usual conditions, might cough up. The crew shouldn't be there.

Moreover, for the particular mountain range where her crew might work, some general patterns are already known. The range has steep drainages. and its rocky cliffs run high above wilderness boundaries. Given the topography and wind conditions, when lightning strikes, wildfires are likely to form a looping pattern: they will start to burn into higher elevations and then, facing thermally driven winds, move back down into the drainage.[7] The up-and-down looping will strengthen while winds spew embers and ignite random spot fires. The resulting hotter temperatures in the drainage will push the flames higher. In a frantic search for fuel, and having nowhere else to get oxygen from, they will jump from live tree to live tree in a fast-moving crown fire. They might even hopscotch over to another ridge. There will never be a "less active" side to manage (by digging a fire line, for instance).

When lightning struck a similar range in June 2016, early in the Northern Rockies fire season, it triggered a major response. Over 600 firefighters were on the scene, as well as a "Type 1" incident management team, 15 bulldozers, 21 engines, 8 aircraft, and 7 water tenders. The fire tripled in size in its first week. It was "an ugly, dirty burn," according to a local fire information officer, who also noted that, owing to "the challenging terrain, the fire has required a lot of aerial resources—it's a big airshow."[8] The information officer pointed to a largely inoperable fire, but that acknowledgment wasn't enough to stop the airshow. Helicopters dropped more than 300,000 gallons of water and 61,000 gallons of retardant on the fire. Air tankers added almost 64,000 gallons of retardant.

Wildfire threats can generally be categorized in terms of fairly stable statistical probabilities. According to the research ecologist Matt Jolly of the Fire Lab, the majority of US wildfires burn under benign conditions; there is a routine aspect in how they are dealt with, and many tend to burn out on their own. For up to 15 percent of wildfires, it is uncertain whether "any change can be effected." And 2 percent of fires will escape control regardless of how many resources are applied. That only 2 or 3 percent of fires "escape" control might sound surprising in a time of multiple wildfires. In fact, since 1985, the number of fires has decreased in the United States. But the acreage burned is going up exponentially, as seen in recent years.[9]

The fire in question eventually died out on its own.[10] It was in the 2 percent range: inoperable. The key is to not disrupt stable probabilities by overintervening. But the system of response did just that: it treated the fire as if it had a chance of being humanly controlled (in the 10–15 percent range). The airshow kicked in. And when it did, it chipped away at stable probabilities by pushing future fires into the 2 percent column; 2 percent becomes 3 percent, and so on—the frequency of extreme fires grows, until the only fire is an inoperable one.

Sometimes letting a fire burn helps keep stable probabilities in check. As one firefighter told me, "fire scars could be beneficial later," preventing future fires from merging or making them burn less severely (because of less accumulated fuel). An inoperable fire allowed to burn might have helped secure an "operable" fire in the hotter and drier near future. The result could have been different if they'd let it burn, he said. It was a lost opportunity to weigh options differently for the climate road ahead: "We would have had a huge barrier to play with for the next twenty years."

The "airshow" is the very antithesis of what this observer wants, which is room to maneuver in the time and space of rapid ecological change, and more horizon. Existing knowledge—in this case, about how a mountain's topography, known fire behavior, and climate change come together to produce extremes—can be schematized and acted upon more effectively to inspire a difficult act of doing "nothing," that is, of standing back or being more selective in the response. For his part, Jolly, who corevised the US National Fire Danger Rating System (for the first time in forty years), believed that the commonality among cases of firefighter fatality today is extreme weather. "And the sad thing," he said, "is that this is very predictable." The new situation requires much more institutional control over where people are and what they're doing in a conscious effort to not expose people unnecessarily in situations "where we know they have almost no probability of success." Someone or something should be held responsible for drawing a line.

The need for that line is as palpable as it is urgent. Beyond it, no amount of resources can match what has been unleashed by decades of fire suppression and global overheating. At these upper limits of bad, the airshow is now an aggressive aerial war on wildfire. This war, partly unavoidable as more communities are threatened, also epitomizes a powerful misalignment between wildfire and its institutional framings—and the fact that sometimes, beneath the airshow, there is no safety zone.[11]

Yet firefighters are still expected to keep diligently working here. As one research scientist told me, "It is the definition of insanity to do something over and over again and expect a different result."[12] For him, "diligent insanity," that is, responding to a changing climate with a subservience to past practices, sets up a future of diminishing options to effect change. What would it take for emergency responders, their agencies, and the public at large to resist reaching this upper limit? The forestry technician might need to refuse the assignment. But this refusal would be out of step with a taxpaying public that expects her and her firefighting crew to go put out the fire.

Beneath the airshow of tankers spewing flame retardant and every tool of suppression available, I met professionals who saw the current necessity for technological overkill as ultimately removing future options to intervene. They pointed to wholesale ecosystem losses like massive forest die-offs due to heat, drought, and infestations; they wondered what mountain, forest, fire scar, vegetation type and now, town, might burn next. The spectacle of the airshow can be read as a symptom of a massive ecological mistake—unleashing a cascade of effects in which hotter fires gut life in such a way "that you don't get the forest back."[13] This loss can lead to other losses and disruptions. Natural infrastructures are also impacted so that, all of a sudden, a water region and its hydrological cycles change, or a sequence of water availability changes: maybe the rivers don't run in September because watershed capture is different, even with the same amount of snowfall.

The year after the airshow brought one of Montana's worst fire seasons of landscape-devouring and infrastructure-eroding wildfires. Records for a lack of precipitation were set as fires burned down to the root channels of dead tree stumps. With exhausted fire crews and widespread smoke pollution and destruction, dispatch services once again hit novel breaking points. In some cases, there was no other choice but to retreat. These developments raise pressing issues for emergency response vis-à-vis climate-accelerated ecological change—namely, about when and how to fight for the future before the circumstances for doing so break away.

Horizon Work in a Time of Runaway Climate Change

A changing global climate system resulting from fossil fuel emissions has fundamentally disrupted our ability to project how the environment will act over time based on established patterns. Human, historical, and climate pressures are changing wildfires, and faltering projections of how they act and how they change are colliding in real time with the dangerous realities of emergency response. Compounded pressures around how to see these changing entities, as well as ingrained assumptions about how nature acts or should act, are making emergency response systems less viable and effective. In this book, I have made a distinction between a world in which natural disasters are contained and our projections hold, and a world in which circumstances refuse to submit to that kind of control.

In world 1, we can delegate emergency response to a group of dedicated professionals, and tomorrow, we hope, will look like today. In world 2, we lose conditions for responsiveness; when the gulf between what is predictable and what occurs grows too large, we risk falling into a state in which only disastrous surprises emerge. "There isn't a ton of horizon," as an interlocutor said to me, and more horizon is what's needed to wrest time from the pace of change and keep spaces of responsiveness together that might otherwise fall apart. Each world implies different takes on the future: in the first, life will go on mostly as usual—disasters being just episodic glitches rather than long-term disturbances that upend life—and, in the second, disaster is accepted as fate. If, in the first, thinking enlists strategically short horizons to buy more time, in the second, thinking carries on with no horizon at all.

Somewhere along the way are the emergency responders, wildfire scientists, fire managers, prescribed burners, and hotshots contending with past experience that is now being rendered obsolete, as well as the policies and legacies that have led to breakaway ecological processes. All the while, as I have shown, they are attempting to build up ways of knowing to keep up with what is being broken. In this effort, they can afford neither denialism nor defeatism and are, rather, engaged in a mode of thinking that considers ecological disasters against a horizon of expectation in which they are still able to act.

In exploring climate futures in terms of horizoning, I shifted the focus from the inevitability of a point of no return that preoccupies so many to the cultural, technical, and otherwise human mediations that hold out the future as recoverable. Just as the climate is changing and expectations for how the environment should act are being shattered, so too are efforts underway to preserve a basic capacity to respond to future shocks. As I have suggested, these efforts involve a responsiveness that does not have the luxury of pattern recognition, implying the need for a rapid reorientation to problems that the scientists, emergency responders, and Indigenous knowledge holders I spoke to see all too well. Yet even as they confront catastrophe, they withstand catastrophic thinking. This stance suffuses scientific paradigms and emergency response systems that, in the face of hard-to-objectify baselines and shifts, can shape broader efforts in configuring livable horizons.

As these pages have shown, runaway change is more than an amalgam of onrushing disasters. Rather, it is the unsettling transformation of things we thought we knew—old fire scars burning, mountains on fire during wet seasons—and the confrontation with forest die-offs and barren landscapes in which nothing familiar returns. The air tanker that increasingly dusts pink flame retardant on communities and homes is a reminder that no amount of technological prowess can match the consequences of what we have unleashed and are now attempting to contain.

For the scientists I got to know, climate change is "pretty much a done deal for the next several hundred years." Acquiring a horizon in the face of this fact means changing cultural expectations around wildfire control. It means revising knowledge calibrated to conditions that no longer exist. It means facing the uncomfortable fact that the circumstances for responding to disaster are themselves disappearing. Acquiring a horizon also entails commitments to safeguarding what hasn't been lost, and recognition of the profound role of returning fire to the land. With models and projections proven less-than-reliable in foreseeing the toll of future disasters, and uncertainty about how to confront rapidly changing ecological phenomena, comes an ethical choice—to provide more knowledge about the fact of uncertainty, or to offer reasonable schemes for action that can make good on always-incomplete information. The actors in this book provide invaluable tools for thinking not only about scientific method, but about the scales of accumulated peril that exist below the perception of our scientific tools.

Rather than capitulating or running toward fear, horizoning is a wayfinding tool that plumbs the lines of a durable world. Those lines run straight through violent legacies, misguided policies, and ongoing structural inequalities. They run through the extractive regimes generating massive transfers of risk that overdetermine who is protected and who is sacrificed in environmental regime shifts. Only in reckoning with these realities can agreed-upon benchmarks for concerted action be imagined, and further damage on a planetary level be stopped. Along the way, it won't be geoengineering, but coordinated acts of stabilization, based in political will and collective responsibility, that can reverse increasingly unlivable planetary conditions and maintain promises of futurity.

There is a deeply personal dimension to horizoning. My interlocutors sought meaning in the mission of suppressing fires and required reasonable norms of safety to do so. At a more basic level, they did not want to see something go away. Love of the woods conspires with a strong sense of duty, as one firefighter told me: "What's crazy is that if the Forest Service says, 'Yep, I want you to go risk your life to save that tree or that road or that trail or that lake or that fishery or whatever it is,' I am okay with that. I have the training and know-how and I believe in the mission. My kids love the woods. I grew up in the woods. I couldn't buy in more to the mission of our national forests and wilderness areas. We're unique in the world. I'll do that. I'll risk my life." This firefighter had deemed the risk to be worthwhile. But he questioned his commitment when it became more than what he bargained for. The public expects emergency responders to act, even when a fire might be one wind gust away from turning into a fireball. The image of the firefighter as hero plays into an illusion that the fireball can be stopped; this image is, as another firefighter told me, referring to his agency, "a facade endorsed by our own people."

Duty is not a matter of blind commitment, but something that needs continued ethical reframing. What do we care for and what do we love? For now, "buying into that hero thing," this firefighter said, perpetuates public misunderstanding: "We're not heroes and we shouldn't be dying like heroes and we shouldn't be memorialized like heroes." Duty shouldn't normalize the outsized environmental costs of fossil-fuel burning that accelerates the global wildfire crisis, or lead to protecting things (homes and private property, often called values) that another firefighter told me he had not signed up to defend: "We're turning into reactive national heroes. Oh, the woods are on fire. And three or four individuals had decided to build a couple half-million-dollar homes in the foothills of California. And now we're going to spend twenty million dollars to save two million dollars. When did our agency become obligated to do that?" Public resources get depleted. "You chose to build your house there. It's insured."

The unwilling "reactive national hero" is asking what it would take for an entire emergency sector—and the rest of us—to acquire a horizon, that is, to rethink automatic acceptance of people like him becoming part of a quasi-militarized operation against nature. The exclusion of fire and the settler colonial legacies and economic mantras underlying this policy are a large part of what has gotten him here. But how human and industrial-scale disruptions now interact with anthropogenic climate change only adds to the uncertainties of where his job takes him, along with many of the first responders fighting wildfires. Under the new conditions we must ask how—or whether—any circumstances for corrective or stabilizing action can be maintained. Are firefighters acting as stabilizers or destabilizers, removing the last ecological safeguards against fire? What if, rather than groundpounding, the real duty is to create ecological futurity? Progress here is measured less in abstract increments of time borrowed or bought, and more in the turning of regime shifts into wanted configurations of the world.

I have been making a case for rethinking the duty amid catastrophe of frontline environmental workers. The public too has an obligation to stop doing the ordinary things that amplify the threats that are now overwhelming them. These workers can no longer deny the problem or help in buying more time. Today, we find ourselves in myriad ways in need of more horizon, which means having the ability to meet conditions of environmental disruption where they are, as well as recognizing our shared responsibility for the catastrophe currently facing frontline workers.

In a well-known speech in 2015, Mark Carney, the governor of the Bank of England, pointed to the cost of climate change denialism for future financial stability, a cost that he referred to as the "tragedy of the horizon." Short-term profiteering off of fossil fuels, Carney argues, will in time be overwhelmed by the long-term ravages of climate change, threatening oil reserve infrastructures and turning fossil fuel assets into "stranded" assets that are unprofitable and "literally unburnable."

In this book, I have considered another kind of stranded asset: human responsiveness itself. As long as salvaging the remainder of fossil fuel reserves holds as a prevailing logic, there is a much more pressing kind of tragedy, one in which the scope of loss exceeds our ability to even imagine or act on it. Will the only remaining image of the world be a world without human presence? In the late 1960s, climate modelers just miles from where I live started performing computer-generated runs into carbon dioxide-saturated worlds on equipment first used for atomic bomb development, simulating global climate change under increased carbon dioxide inputs. When Syukuro Manabe, a climatologist and modeling pioneer, pumped CO_2 concentrations of four to six times present-day concentrations into his model, he found that the resulting high temperatures approached "the days of [the] Cretaceous period," a super-greenhouse world from a hundred million years ago.[1]

The physicist James Hansen sounded the alarm on the looming climate crisis in his 1988 congressional testimony. Two decades earlier, he had analyzed the planet Venus and its surrounding atmosphere. As the physicist parsed the molecules and dust of a planet twenty-five million miles from Earth, he noticed something peculiar: it was made of carbon dioxide gas (97 percent). He had to wonder whether, swallowed up by the horizons that surround it, Venus was an artifact of some runaway process, a relic of a planet that once looked like Earth, billions of years ago. Perhaps lifeless Venus had been forced into some ghastly chemical oblivion. But this image was no match for a more hopeful one: the iconic "blue marble" photograph of Earth, taken in 1972 at a distance of twenty-eight thousand miles by moon-bound crew members of the Apollo 17 spacecraft. Galvanizing environmental movements in the early seventies, the photograph captured Earth's teeming vitality in oceans of presumably dead planets. The difference between the two sheds light on the unexpected space between life and death that is more like a question: How do we circumvent inevitability, or, rather, how do we protect favorable circumstances within the vertical emissions curve that could prevent us from being carried away into extinction?

This juxtaposition of life and death takes me back to where this book began: a small border town where bombs fell and buried refugees alive in a makeshift bomb shelter. The surviving refugees witnessed the bodies piled on a flatbed truck, and one sister's scream enabled Misio, presumed to be dead, to regain consciousness and get off the truck. By that time, he had already practiced his ability to find safe distance from where other bombs fell. My childhood preoccupation with that distance, I now realize, is about the cruel imprecisions of extinction boundaries themselves.

The struggle to horizon, then, is about not only achieving better projections of risk or uncertainty, but cutting a different path, carving out effective perceptual range, and making futures less remote. It is a response to jeopardy faced by entire systems and an effort to recover the circumstances in which such cognizance becomes actionable rather than obsolete. When destruction obliges us to revise knowledge calibrated to conditions that no longer exist, our marking of horizons beyond which the world as we know it disappears is itself an exercise in delimiting the knowable, and thus habitable, world.

ACKNOWLEDGMENTS

This work would not have been possible without the many people who inspired it. I am deeply grateful to the ecological theorists and experimentalists, wildfire scientists, emergency responders, and other practitioners and thinkers who shared their time, insights, and experiences with me. Their guidance and creativity mark out needed directions and show how inclusive ecologies of knowledge are crucial now and for the future. My sense of gratitude also extends to people I have advised past and present, and who in their lives as teachers, scholars, and physician-scholars are nurturing equitable ecologies of research and care across disciplines in the social sciences, clinical medicine, and beyond. Thank you, Michael Joiner, with whom enriching conversations over the years have pushed my thinking and made this work always worth pursuing. I also am especially grateful to Naomi Zucker, whose thought-provoking comments on the lives and conditions it reflects were invaluable and too numerous to count; and to Cameron Brinitzer, whose clarity and careful readings were critical to my composing of initial drafts and making essential connections. Over the years, I have learned from colleagues, mentors, and neighbors who have offered feedback in key moments in the development of this project, including Michael Fischer, whose generosity and openness to ideas creates more light and interpretive possibility for so many, as well as people who on different occasions and in various moments of this work provided important feedback and intellectual uplift. Thank you to Arcadio Díaz-Quiñones, Arthur Kleinman, Melissa Lane, Deborah Winslow, Sharon Kaufman, Diane Nelson, Candis Callison, Rob Nixon, Anne McClintock, Bethany Wiggin, the late Paul Rabinow, James Boon, Jean Comaroff, Limor Samimian-Darash, George Marcus, Jessica Cooper, Rosanna Dent, James Faubion, Dominic Boyer, Cymene Howe, Beto Verissimo, Tasso Azevedo, Marcel LaFlamme, Colin Hoag, Lucas Bessire, David Bond, Joseph Amon, Marcus Brinitzer, Patricia Shanley, Christopher Barr, Susann Wilkinson, Arbel Griner, Peter Locke, Helena Hansen, Puneet Sahota, Fran Barg, Onur Gunay, Paulo Barreto, Tim Neale, Renan Moura, Dorothea von Moltke, Susan Reynolds Whyte, Michael Whyte, Jens Seeberg, Tali Ziv, Lotte Meinert, Guilherme Fagundes, and Andreas Roepstorff, as well as Kevin Burke, Paul Mitchell, Sharon Jacobs, Rebecca Mueller, and the late Timothy Powell. In the MD-PhD Program in Anthropology that I have had the privilege to direct, I would like to extend my special thanks to Caroline Hodge, Josh Franklin, Randall Burson, Alex Chen, Michelle Munyikwa, Benjamin Sieff, Angela Ross Perfetti, Ankita Reddy, and Nipun Kottage, physician-anthropologists who have been an inspiration to mentor. For motivating conversations and ever astute listening moving ideas along, I thank Joseph Lee Young and Utpal Sandesara. Sara Rendell's care and attention to the effervescent details that make life possible has enriched this work. My gratitude goes to Mike Levine and Carolyn Bond for critical editorial feedback. Eric Henney's and Ken Wissoker's commitments to ideas are models of engaged social science scholarship. The works of Stephen Pyne, Robin Wall Kimmerer, and Kyle Powys Whyte, among others, are as crucial as ever, as are the far-reaching scholarly publications and educational media of the CSKT Séliš-Qlispé Culture Committee.

At the Department of Anthropology at the University of Pennsylvania, I am grateful to staff and colleagues whose work and thought enrich me daily, including Kathy Morrison, Nikhil Anand, Kristina Lyons, Andrew Carruthers, Marge Bruchac, Deborah Thomas, Lynn Meskell, Lauren Ristvet, Tad Schurr, and Greg Urban, as well as colleagues in the History and Sociology of Science, particularly Robert Aronowitz, Susan Lindee, and Ramah McKay. Engagements of students over the years, including Ivana Kohut, Adriana Purcell, Omar Husni, Sam Warrick, Anand Muthusamy, Jessie Lu, and Isabelle Breier, have been helpful in many ways. My teaching and advising take place on the traditional and contemporary homelands of the Lenape, known as Lenapahoking, and whose descendents include the Delaware Tribe and Delaware Nation of Oklahoma; the Nanticoke Lenni-Lenape, Ramapough Lenape, and Powhatan Renape of New Jersey; and the Munsee Delaware of Ontario.

The seeds of this project were first presented at Princeton University's Anthropology of Becoming seminar and at Aarhus University. I have benefited from exchanges with colleagues and students where pieces of the project were presented, including at New York University, University of Oslo, Northwestern University, School for Advanced Research, University of California, Irvine, The Graduate Center, CUNY, Rice University, Harvard University, and Johns Hopkins University. An earlier sabbatical from the School of Arts and Sciences at the University of Pennsylvania, followed by a productive summer at the School for Advanced Research (SAR), helped me commence research for this book, as did support from the National Science Foundation, grant #1646822. I am grateful to Princeton's University Center for Human Values and its staff for an ideal year of critical thinking and writing. The generosity of the John Simon Guggenheim Foundation allowed me to complete the manuscript. Any conclusions, findings, or recommendations are my own and not necessarily those of sponsoring agencies.

My very special thanks goes to Christie Henry, director of Princeton University Press, for her wisdom and support of this project. Fred Appel has been a wonderful editor and supporter of this book, giving editorial guidance at key junctures. I owe a special debt to the two reviewers for their generous and sharp comments. I was lucky and am grateful to be working again with Lauren Lepow, production editor, as well as with Lisa Black, Dimitri Karetnikov, Layla MacRory, Terri O'Prey, and indexer Sylvia Coates. For their constancy and care throughout, I thank Helene Katz and Peter Yi, as well as Adrienne Jensen, Alison Brancone, and Hilary Friedman.

I am so grateful to my family, especially to Andre Biehl for insight, clarity, and the immense joy you bring. My mother, Tania Petryna, constantly teaches me about generosity and how to truly lift up others. To Michael Petryna, whose last words, see you in the horizon, carry across these pages. For their care and presence, I thank Noemia, Vera and Newton, Oksana and Andriy, Christina, Mark, Renee, Ron, and Luba. João Biehl's love and integrity as a human being and life partner, like the numerous stars in the sky, inspires me throughout.

NOTES

Chapter 1. What Is the Upper Limit?

1. This number typically refers to outdoor levels of CO_2. In general, indoor levels are higher.

2. On addressing these psychological barriers as "dragons of inaction," see Gifford 2011:291.

3. Holcomb 1916:445.

4. Barash et al. 2019:11.

5. See Hwang et al. 2013.

6. See Cannon 1942. It refers to the "dramatic suddenness of the illness following [a] threat, coupled with a lack of any apparent injury, exposure to toxins, or infection." For Cannon it suggested "that merely the fear of death could, through physiological response mechanisms initiated by fear, precipitate death itself" (Sternberg 2002:1564).

7. Cannon 1915:205.

8. See Telch et al. 2012. On CO_2 emissions and impairments in human cognitive response, see Karnauskas, Miller, and Schapiro 2020.

9. See Troisi 1957 and "Silo Gases—the Hidden Danger," Penn State Extension, for example.

10. Epler 1989:368.

11. Hayhurst and Scott 1914:1570,1571.

12. For animal welfare advocates' preferences, see Grandin and Smith 2004.

13. On poor ventilation and overcrowded shelters, see Pryor and Yuill 1966:68–69.

14. Caidin 1960, cited in Lucas et al. 1990:813. See also Bond 1946. The writer Kurt Vonnegut survived the Dresden air raid in a bomb shelter as a prisoner of war and also describes his experiences in *Slaughterhouse 5*. On the incompleteness of these postmortem accounts, see Sebald 2011.

15. Either a landslide, an earthquake, or a small volcanic eruption inside the lake triggered the explosion from CO_2 gas that was trapped in the bottom of the lake. Crew 2017.

16. Recollection of Lake Nyos disaster survivor Joseph Nkwain, Taylor 2011:167.

17. CO_2 makes up about 0.04 percent of air.

18. According to pathology reports, injuries like pressure sores were ascribed to the heat release, but the only "heat-related" injuries were those from people falling into cooking fires (George Kling, personal communication). On degassing steps, see Halbwachs, Sabroux, and Kayser 2019 and Kusakabe et al. 2008.

19. Sebald 2011:4, 78.

20. Forthofer and Goodrick 2011:3, cited from Ebert 1963. Also see Lucas et al. 1990.

21. Sebald 2011:27. On Allied bombings as weapons of mass destruction, see Friedrich 2006, Lindqvist 2000.

22. Sullivan 2018:2.

23. Lucas et al. 1990 argues that nuclear bomb shelters would protect against firestorms. On the theme of variable and massive fire effects being excluded from damage assessment models to promote images of a "limited and 'winnable' nuclear war," see Eden 2006. On disease preparedness, see Lakoff 2017.

24. See James and Macatangay n.d. Some of the researchers, funded by the US Public Health Service, were anesthesiologists (Brechner et al. 1965).

25. Mbembe 2020. The translation from French has been slightly modified. To "keep breath in the Black body" amid anti-Black racism, see Sharpe 2016:109.

26. Wallace-Wells 2019a; 2019b.

27. On "multipliers" of threat known collectively as abrupt climate change, see National Research Council 2013. The "slow violence" of climate change, as Nixon outlined in his seminal book (2011), and its repercussions in postcolonial and global injustices, have been by compounded by these manifold threats.

28. Shukman 2013; National Geographic 2019.

29. This system varies with "salinity, temperature, carbonation, photosynthesis and eventually ionic exchange in the gills. In no two waters is the pattern of CO_2 exchange the same." See Dejours 1978. Also see Ishimatsu et al. 2005.

Chapter 2. Building Perceptual Range

1. Forbes 1887 (my italics).

2. See Ault 2016, Povoledo 2017, also Yamauchi et al. 2014.

3. The themes of time and futurity take on meaningful dimension in anthropologies of climate change, including on carbon markets and their significance (Whitington 2020); the contours of renewable energy markets (Günel 2021); wind park development and Indigenous communities' resistance to bureaucracy and corporate power (Howe 2019; Boyer 2019). Scholars have also addressed knowledge production in strange weather "experiments" (Zee 2017) and in climate change as an emergent form of life (Callison 2014), while also challenging the Anthropocene's unitary planetary time frame (see Moore 2016). For analysis of temporal framing as it remakes action within unstable social systems, see Fischer 2018.

4. Singer 2014. On rules of cooperative survival, see, for example, Hamilton's rule, which genetically and mathematically defines the horizons of an individual actor's selfishness: http://brembs.net/hamilton/.

5. Cahill et al. 2013; Griffen and Drake 2009; Pimm 2009.

6. *Britannica Guide to Climate Change* 2008.

7. See Pimm 2009:R600; Cahill et al. 2013:2.

8. Contrast this myopia with fish that "can choose when to be caught" as agential beings (Todd 2014:222). Also see Sheila Watt-Cloutier (2020), who notes how Inuit hunters in Nunavik "were obliged to kill only animals who 'presented' themselves for the taking" (Watt-Cloutier 2020: 272). For a consideration of Indigenous stewardship of fire as an agential being, see chapter 6.

9. Forbes 1887.

10. On the notion of settler time, see Rifkin 2017. On how colonial and ecological violence stand in relation to one another, see Bonilla and LeBrón 2019. On "the anthropo-not-seen" as "the will to end many worlds that produced the one-world," see De la Cadena 2015:3. On how the impulse to "return to the presettlement equilibrium" repeats the erasure of Indigenous inhabitants' knowledge and relations to the lands and waters of these sites, see Kimmerer 2000:5. The University of Wisconsin–Madison recently recognized campus land as the ancestral home of the Ho-Chunk. Erickson 2019. On the limits of this reckoning, see Greendeer 2019.

11. Dunning et al. 1884:443. Some of Forbes's own histological samples were destroyed; he was "unable to obtain good material enough from which to generalize" (Forbes 1887:487).

12. Dunning et al. 1884:442.

13. "Birds and fish dying. Strange and unexplained mortality in Wisconsin," *New York Times*, August 11, 1884.

14. Nietzsche 1909. According to a different translation, this passage reads as "it will wither away feebly or overhastily to its early demise." Nietzsche's subjects perish through isolation (that comes from having the "narrowest of horizons . . . as narrow as that of an Alpine valley"). Nietzsche 1997. The theme of horizons arises in philosophy, literary theory, and the social sciences. Hans Robert Jauss coined the term "horizon of expectations" to refer to larger contexts of meaning in which readers and authors meet in a literary experience (1982:22). He derived insight from Hans-Georg Gadamer (1997), who writes of that meeting ground as a "fusion of horizons" transforming both. In theorizations of queer horizons, José Esteban Muñoz writes of a "modality of ecstatic time in which the temporal stranglehold that I describe as straight time is interrupted or stepped out of" (2009:32). On imaginative horizons as a "dialectic between openness and closure," see Crapanzano 2003. On abundance and a "scarcity of social futures" in a digital era, see Appadurai 2021. On turning nouns into verbs, and on anthimeria as a poetic technique in which nouns become verb forms, see Brogan 2012.

15. Ibid. Here the pronoun "we" refers to humans as a zoological species as posited in eighteenth-century Western taxonomy, which excluded humans not recognized as fully human by this very same taxonomy. Wynter 2003, Weheliye 2014.

16. Napier 2014, cited in Petryna and Mitchell 2017.

17. See Hughes et al. 2012; Griffen and Drake 2009.

18. Einhorn et al. 2020.

19. Carpenter 2003. Also see Scheffer et al. 2012, Carpenter et al. 2011, Lenton 2011, Lenton et al. 2008.

20. Lubchenco and Hayes 2012.

21. See NOAA 2021. Records set in 2011 were surpassed in 2017, when the cumulative costs of weather and climate disaster events exceeded $300 billion in the United States.

22. Lubchenco and Hayes 2012.

23. National Research Council 2013:27, 80.

24. Solomon et al. 2008.

25. Pizer noted these damage values as "most challenging" to estimate (2017:1330). On routine underestimations of damage, see Ahmann and Kenner 2020, for example.

26. Lin and Petersen 2013.

27. One example relates to the disappearance of Earth's so-called "carbon sinks." While most CO_2 goes into the atmosphere, land, and ocean, researchers are not sure where the inordinately excessive CO_2 will go. See Lahsen 2009.

28. White 2014.

29. Hughes et al. 2018:81. "Earth's delicate web of life" quoted from Berwyn 2020.

30. Krenak 2020.

31. Kopenawa 2013:29, 32.

32. Harjo 2019:25. On futurity as creating "the 'open, future, possible' by probing the human ability to act," see Eshel 2012:4.

Chapter 3. When Paths Disappear

1. *Chasing Ice* 2013. The ice retreated farther between 2001 and 2010 than it had in the previous hundred years.

2. NASA 2020.

3. Mooney 2020.

4. On cloud loss, see Wolchover 2019. On interconnected crises and "fugue times," see McClintock 2020.

5. Guterl 2012:4. On emissions rising in 2021, see IEA 2021. To hold temperature rise to the 1.5 degree Celsius target, cutting emissions of methane, an extremely consequential greenhouse gas, is as urgent as cutting carbon dioxide emissions.

6. Mahony and Hulme 2012:78. See also Randalls 2015. On calculation and distillation, see Ballestero 2019.

7. Socolow 2020:48.

8. Socolow 2011.

9. On the meanings of geoengineering for a climate-changed world, see Kolbert 2021.

10. Davis et al. 2013.

11. Marcia McNutt, public lecture, National Council for Science and the Environment, Disasters and Environment Conference, Washington, DC, January 2013.

12. Ibid.

13. On switches. see Nordenson 2016. On Germany's floods, see Eddy et al. 2021.

14. Magnuson helped establish the US National Science Foundation–funded Long-Term Ecological Research (LTER) Program. Originally involving eleven sites spanning

distinct biomes, the program has twenty-eight sites where more than eighteen hundred scientists and students conduct long-term ecological observation, experiments, and modeling.

15. Magnuson 1990:495, 497.

16. Magnuson 1995:454. The quote continues: "biological relics persist even after conditions change, . . . and a chain of events accumulates the lags between cause-and-effect events."

17. Magnuson 1990:495. For a relevant typology of disaggregated natures past tipping points and their ethical framing, see Fischer (2009), which takes up the call to action by Meyer (2006).

18. Magnuson, Bowser, and Beckel 1983. The LTER program's initial planners (see above note) placed special priority on obtaining baseline references for the "appropriate study of the patterns of disturbances and responses to disturbances." The sites selected "would be those subject to only minimal influence by human activity" (LTER Network 1979). Indigenous communities were not involved in constructing baseline references. To identify the Indigenous lands on which an LTER site is located, consult Native-Land.ca.

19. On spurious certainty, see Biggs, Carpenter, and Brock 2009.

20. On the "stability of ecological systems," see Holling 1973. On "structural stability and morphogenesis," see Thom 1975.

21. See Steffen et al. 2018:8256.

22. Lake Mendota is one of several Wisconsin lakes constituting the North Temperate Lakes LTER site, which Carpenter led from 1999 to 2009.

23. Not only the community structure changes, but the gas exchange with the atmosphere, the processing of nitrogen and phosphorous, the sedimentation—"all of these things change. And this dynamic is completely unexplainable through the traditional linear models of the world." In this vein, Holling's (1998) insight about competing hypotheses and building an "increasingly credible line of tested argument" in ecological research holds true: "Not only is the science incomplete, the system itself is a moving target, evolving because of the impacts of management and the progressive expansion of the scale of human influences on the planet."

24. Stephen Carpenter, quoted in Zagorski 2005.

Chapter 4. Horizon Work

1. Thomas 2004:21.

2. Parunak et al. 2008.

3. In commenting on producing reliable observations of Mars, one astronomer notes, "Calibration is essential for any instrument you send into space. You're going into an unknown environment, measuring things that no one has ever encountered before. So how do you know you can trust what your instrument's telling you? . . . [W]ithout [calibration] we'd never be able to figure out what our readings on Mars

meant" (Squyres 2005, quoted in Vertesi 2015:54). For a wonderful analysis of the production of trustworthy images of Mars, see Vertesi 2015.

4. On "scaling rules," see Griffen and Drake 2009; on "maintaining a safe distance from dangerous thresholds," see Rockström et al. 2009, cited in Hughes et al. 2012:153.

5. Gladwin 1970:232. This coral atoll was formerly known as Puluwat.

6. On autonomous fault detection and system repair in space rovers, see Verma, Langford, and Simmons 2001. On dead reckoning as a transformation of invisible presents into effective perceptual range in seafaring, see Huth 2013. On the uses of horizons to "expand the target" in navigation, see Lewis 1971. Verma, Langford, and Simmons 2001.

7. Scheffer 2009.

8. Rocha et al. 2018:1379.

9. Wang et al. 2012.

10. Coleman 1964.

11. Mayr 1982.

12. Thompson 1942:1094–1095.

13. Gould 1971:250. Second quote from Thompson 1942:7.

14. Gould 1971:253.

15. See Waddington 1942.

16. These homely phenomena included "the cracks in an old wall, the shape of a cloud, the path of a falling leaf, or the froth on a pint of beer"—"in themselves trivial (often to the point that they escape attention altogether!" (Thom 1975:8). In 1962, Thom encountered writings on "morphogenetic fields" in embryology by C. H. Waddington, with whom he would collaborate and publish. See Petryna and Mitchell 2017.

17. In the early seventies, Thom's influence peaked then waned amid criticism of his theory being applied too widely. The related term, tipping point, first appeared in research that tracked racial segregation of neighborhoods in the United States (Edsell 2015). It was then popularized by Malcolm Gladwell in his 2000 book by the same name. Climate scientists had adopted the term to account for nonlinear ecosystem behaviors.

18. Thank you to Michael Joiner for these insights.

19. See Steffen et al. 2018. This analysis points to processes that "could push the Earth System toward a planetary threshold that, if crossed, could prevent stabilization of the climate at intermediate temperature rises and cause continued warming on a 'Hothouse Earth' pathway even as human emissions are reduced." Within a decade of the publication of "Planetary Boundaries: Exploring the Safe Operating Space for Humanity" (Rockström et al. 2009), the nature of the analysis around planetary life-support systems has changed. No longer is it recognizing and "maintaining a safe distance from dangerous thresholds"; it is now a race to stabilize entire ecosystems.

20. Lovejoy and Nobre 2018. On "convincingly established" tipping points, see Lenton et al. 2008:1786.

21. Beto Verrissimo, cited in *Amazonia Undercover* (film). Ciavatta 2019.

22. Brysse et al. 2013:330.

23. See Kolbert 2021; Barrett et al. 2014; Sapinski et al. 2020. A key issue concerns scalability, particularly in how projects are assumed to become larger "without changing the nature of the project"; see Tsing 2012:507 and Tsing 2015.

24. Newburger 2019.

25. Data from the European Union's Copernicus Atmosphere Monitoring Service https://atmosphere.copernicus.eu/wildfires-continue-rage-australia.

26. Ferguson, Sekula, and Szabó 2020. Also see Barreto et al. 2017.

27. Azevedo and Sizer 2019. See the pivotal work of Imazon (https://imazon.org.br/en/) and MapBiomas. On ethics and "irreversible" crisis, see Rojas 2016.

28. On the Anthropocene as Pyrocene, an age of fire as consequential as the last ice age, see Pyne 2021.

Chapter 5. "Throw Away Your Mental Slides"

1. Kahneman and Klein 2009.

2. Such intuitive expertise evolves in relatively stable environments, as the authors above suggest (ibid.). On expertise, risk, and knowledge production, see Jasanoff 2016, Dumit 2014, Fortun 2001, Knowles 2011, and Carr 2010, among others.

3. Holthaus 2017. The term "megafire" has no single definition, but it typically describes fires over 100,000 acres. On the role of anthropogenic climate change on wildfire across western US forests, see Abatzoglou and Williams 2016.

4. Downey 2017.

5. For brevity, I use the term "fire manager" instead of "fire management officer."

6. Migliozzi et al. 2020.

7. Such as those of 1910 and 1988. See Pyne 2016.

8. It burned three million acres across the western United States and Canada. See Pyne 2001.

9. "Type 1" incident management teams engage large-scale, complex incidents that include a variety of disasters, such as hurricanes, floods, fires, earthquakes, and tornadoes.

10. National Park Service, USDA Forest Service, "Fire Terminology," https://www.fs.fed.us/nwacfire/home/terminology.html#I.

11. Wildfire response in the United States occurs within a federal, state, and tribal interagency environment. This number, thirty-two thousand, includes personnel drawn from the US Forest Service, the Bureau of Land Management, the US Fish and Wildlife Service, the Bureau of Indian Affairs, and the National Park Service. Wildland firefighters are drawn from private contracting companies, the military (e.g., the National Guard), prison-based crews, and local fire department volunteers. As large wildfires become more frequent, "the number of personnel involved in wildland fire suppression will continue to grow." See Butler et al. 2017. Women constitute 12 percent of wildland firefighters, working in different positions and career tracks. Gender discrimination and harassment persist (Flock and Barajas

2018, Langlois 2017). There are all-female inmate crews and fire crews in the Forest Service and Conservation Corps.

12. Ricardo Garcia, letter to the editor, *Santa Fe Reporter*, October 23, 2019. Also see Hay 2019.

13. Crown fires occur in a range of inoperability (currently estimated as 2 percent of fires occurring in the United States). Also see Jolly et al. 2015.

14. Both quotes are from Curwen 2017.

15. On response-ability, see Haraway 2016:35.

16. The Cerro Grande Fire was an escaped prescribed fire. Armstrong was one of several local Forest Service officials who advocated for more prescribed burning to mitigate fuel loads before this fire, which burned near the Los Alamos Laboratory. Armstrong's efforts reduced damage to surrounding neighborhoods (Ribe 2010:345).

17. J. Thomas 2020. On the politics of nature in New Mexico. see Kosek 2006.

18. Morton 2013:1.

19. Irfan 2020. On gradients, see DeLanda 2010.

20. Short 2017:33. Also see Finney et al. 2009.

21. Grinberg, McLaughlin, and Zdanowicz 2018.

22. Small trees and shrubs next to destroyed structures "most likely ignited and burned from the already burning structures" and the fire destruction, he said, "was due to the ignition characteristics of the structures (including flammable debris)."

Chapter 6. "You Can't Take Fire Away"

1. Agee and Skinner 2005:84.

2. Spence 2017.

3. Archaeologist Matt Liebmann and noted tree-ring researcher Thomas Swetnam have studied Native American population decline between 1492 and 1900 CE instigated by the European colonization of the Americas. Particular work took place in the Jemez Province of New Mexico. See Liebmann et al. 2016:E696.

4. Settler colonialism is a form of domination in which Indigenous peoples are erased through genocide, forced removal, cultural destruction, and repression. Settler colonial erasure is an ongoing process that "normalizes the continuous settler occupation, exploiting lands and resources to which Indigenous peoples have genealogical relationships." See Cox 2017.

5. Agee and Skinner 2005:84.

6. Whyte 2018a and Gilpin 2019. Anthropogenic inflictions of extinction are reflected in what White Earth Ojibwe historian Jean M. O'Brien (2010) calls "firsting and lasting"—in which European Americans assert realities, in this case about climate change impacts, that deny history to Indigenous peoples. O'Brien 2010. On how colonialism inflicted various forms of anthropogenic climate change, see Reo and Parker 2014.

7. Harjo 2019: 33,13. On wayfinding as interpretive craft, see Ingold 2011.

8. "Frontline communities are those that experience climate change first and often feel the worst effects. They are communities that have higher exposures, are more sensitive, and are less able to adapt to climate change impacts for a variety of reasons."

"Indigenous Peoples Terminology for the Fourth National Climate Assessment," http://www7.nau.edu/itep/main/tcc/docs/resources/Indigenous%20Peoples%20 Terminology%20for%20NCA4_final.pdf. On the place of Indigeneity in geoengineering discourses, see Whyte 2018b. On the ethics of naming and linking climate change to destructive cultural, political, and environmental changes, see Callison 2014.

9. Whyte 2018a, 2017.

10. The Flathead Indian Reservation, located in western Montana, is home to the Bitterroot Salish, the Upper Pend d'Oreille, and the Kootenai Tribes. All of western Montana, as well as parts of Idaho, Wyoming, and British Columbia, constitute the territories of these tribal communities, amounting to over 20 million acres. The Tribes ceded most of this territory with the Hellgate Treaty, signed in 1855. The 1.3 million acres that remained became the Flathead Reservation. When land allotment began in 1904, over 500,000 acres were taken from tribal ownership. "CSKT Characteristics and History of the Tribes," *Climate Change Strategic Plan* 2103:4.

11. See Confederated Salish and Kootenai Tribes 2005. Also see *Fire on the Land*, https://fwrconline.csktnrd.org/Fire/FireOnTheLand/History/TraditionalCulture/.

12. On "interaction with the natural world," see CSKT Climate Change Strategic Plan 2013:6. On the "moral covenant of reciprocity," see Kimmerer 2013. On the "gift of fire," see White 2007.

13. Confederated Salish and Kootenai Tribes 2005. On integrating Indigenous knowledge with Western science to address wildfires and forest health, see Mason et al. 2012. Scholars have made critical contributions to our understanding of models of fire management as they relate to Indigenous fire expertise. On fire suppression as a form of colonial violence and the growth of Indigenous fire science and management among the Karuk people in California, see Norgaard 2019. On critical collaborations between Indigenous knowledge holders and settler scientists in a changing climate in Australia, see Neale et al. 2019; Verran 2002. On fire management among *quilombola* communities in Brazil, see Fagundes 2019. Also see Fowler and Welch 2018.

14. Salish-Pend d'Oreille Culture Committee and Elders Cultural Advisory Council, (CSKT) 2019:31.

15. Evidence shows continuous occupancy in western Montana "reaching back to the end of the last Ice Age." Ibid. 2019:9.

16. CSKT, *Climate Change Strategic Plan* 2013:iii. On guidelines for integrating traditional knowledges in climate change initiatives, see Climate and Traditional Knowledges Workgroup 2014.

17. To learn more, consult "The Séliš-Q̓lispé Ethnogeography Project" 2019a; The Salish-Pend d'Oreille Culture Committee and Elders Cultural Advisory Council (CSKT) 2019; The Séliš-Q̓lispe Culture Committee 2019b; and Weber 2017.

18. Salish-Pend d'Oreille Culture Committee and Elders Cultural Advisory Council (CSKT) 2019:24.

19. Confederated Salish and Kootenai Tribes 2005.

20. Callison 2020:135. On the meaning of stability of place for plant and animal resources, see Salish-Kootenai Fire History Project 2006. Given the ongoing

assessments that Callison writes about, a place-name not so much about "fashioning a place-world" where naming and geographic features unite, as Keith Basso wrote in his study of place-names in Apache life and landscapes (see Basso 1996:11). For a critique of Basso as not sufficiently addressing "how colonial spatial restructuring of land" suppresses the "voices" of the land and its specific relationships, see Goeman 2008.

21. Todd 2018:65. Watts employs Anishinaabe and Haudenosaunee philosophy to build the concept of Place-Thought (2013). Davis and Todd (2017) show how a concept of integrated "land and thought" (769) is erased in settler colonial framings of the Anthropocene.

22. Watts 2013:33.

23. Ritchie 2019.

24. See Neale (2020), which reflects on Whyte's notion of Indigenous knowledge practices having governance, not mere supplemental, value.

25. Swetnam et al. 2016.

26. Trahan, CSKT fire technician who employs prevention burns. Spence 2017.

27. See Elders and Fire Managers Interviews, *Fire on the Land: Native People and Fire in the Northern Rockies*, http://fwrconline.csktnrd.org/Fire/index.html.

28. Jamison 2005.

29. Kimmerer 2013:363.

30. White 2007. First contact occurred in 1805, when the Lewis and Clark expedition encountered the Salish in the Bitterroot Valley of Montana. On myths of pristine landscapes, see Cronon 1995, for example.

31. Swetnam et al. 2016.

32. On Indigenous fire stewardship as "biophysical stewardship," see Lake 2021:31.

33. CSKT, *Climate Change Strategic Plan* 2013:38. For seminal work on loss and moral imagination on the Flathead Reservation, see O'Nell 1996.

34. Chief Charlo, April 26, 1876. "Indian Taxation," *Weekly Missoulian*, cited in Bigart and McDonald 2020:37–40.

35. White, interviewed for Montana Public Radio's Fireline podcast (2021). https://www.firelinepodcast.org/episode-4-the-gift-of-fire/.

36. Salish-Pend d'Oreille Culture Committee and Elders Cultural Advisory Council, (CSKT) 2008.

37. Agee and Skinner 2005:84, citing Pyne 2001. On the history of the Big Burn, see Egan 2009 and Pyne 1982, 2001.

38. Diamond 2011:44.

39. Tony Harwood, personal communication, 2018.

40. He was part of a team of interdisciplinary specialists who developed the CSKT Forest Management Plan. To provide long-term guidance, the plan took an ecosystem management approach: it combined ecological and traditional knowledge and other scientific, economic, and managerial principles to create a plan of action that continues to inform the Tribes' forest and fire management activity.

41. A smokejumper is a wildland firefighter who jumps (with a parachute) into a forest fire to make what is called an initial attack. Interagency partners include the

Bureau of Land Management, US Fish and Wildlife Service, national forests, the Nature Conservancy, and state, county, and rural agencies.

42. Swaney 2021.

43. Ibid. On barriers to expanding prescribed fire and recommended solutions, see Clark et al. 2020.

44. On camas reappearing in other areas after fires, see Tony Incashola Sr and Jr, interviewed for Montana Public Radio's Fireline podcast (2021). Also see Swaney 2021. On abundance of woods potato, woods onion, things that were traditionally gathered in areas that would not be there without fire, see Harwood 2021.

Chapter 7. Witnessing Professionals

1. Arendt, "The Gap between Past and Future," 1961:6.

2. Hotshot Superintendent, quoted in Canon 2020.

3. Moyers 2017.

4. Ibid.

5. Lifton 2017.

6. Gottbrath 2020.

7. Timothy Ingalsbee, Yale Wildland Firefighter Rights Initiative—COVID-19 Roundtable, August 2, 2020. On fire-suppression resource scarcity, see Belval, Stonesifer, and Calkin 2020.

8. On a critique of the values-at-risk concept, see Zahara 2020.

9. Rempfer 2018.

10. As of 2020, there were 113 hotshot crews in the US. A "Type 1" crew requires the greatest amount of personnel and equipment; a "Type 5" would require the fewest resources. The designations for wildfire severity range from a Type 1 to a Type 5 incident within the Incident Command System (ICS). See National Park Service, "Wildland Fire: Incident Command System Levels," https://www.nps.gov/articles/wildland-fire-incident-command-system-levels.htm.

11. US Forest Service, "Hotshots: These Handcrews Can Really Take the Heat!" https://www.fs.usda.gov/science-technology/fire/people/hotshots.

12. Timothy Ingalsbee, Yale Wildland Firefighter Rights Initiative—COVID-19 Roundtable, August 2, 2020.

13. Casey Judd, ibid. Proposed legislation aims to secure expanded workers' compensation benefits and COVID-19 testing (see S.3910: The COVID-19 as a Presumptive Disease in Wildland Firefighters Act).

14. In many countries, wildfires are fought by volunteer firefighters; response systems have not evolved in the direction that US fire management has.

15. Lavender 2013.

16. On a US bill to raise the raise the maximum limit on overtime pay for federal firefighters, see Gabbert 2021.

17. Short 2017:33. On more effective wildland fire management, see Thompson et al. 2018. On "suppression effectiveness knowledge gaps," see Plucinski 2019.

18. The National Environmental Policy Act (1970) requires managers to assess environmental impacts prior to the building of "airports, buildings, military complexes, highways, parkland purchases, and other federal activities [that] are proposed." See EPA, "Summary of the National Environmental Policy Act," https://www.epa.gov/laws-regulations/summary-national-environmental-policy-act.

19. *Forest Serv. Emps. for Envtl. Ethics v. U.S. Forest Serv.*, No. 17-35569 (9th Cir. Jun. 8, 2018). https://files.acrobat.com/a/preview/384a0759-6528-4e26-bd9d-f444e1750dde.

20. Ibid.

21. Forest Service Employees for Environmental Ethics (FSEEE), Preparing for "Emergencies," https://www.fseee.org/2018/02/06/preparing-for-emergencies/.

22. See *Forest Serv. Emps. for Envtl. Ethics v. U.S. Forest Serv.*, No. 17-35569 (9th Cir. Jun. 8, 2018).

23. See Forest Service Employees for Environmental Ethics, https://www.fseee.org/victories/. The federal district court judge would maintain that forest fires are emergencies.

24. US Forest Service, "Wildland Fire," https://www.fs.fed.us/fire/people/handcrews/about_handcrews.html.

25. Puerini and Torres 2020.

Chapter 8. "Waiting for a Reality Response"

1. For example, the work of fire-behavior scientist Mark Finney, quoted here, aligns theory, basic science, and practical modeling to advance fire management objectives.

2. In their classic *Leviathan and the Air-Pump*, Steven Shapin and Simon Schaffer (1985) show how a vacuum is not just an empty space; it is a political and technical achievement in which truths depend on the integrity of their instruments as much as on processes of elimination. I draw the phrase "what we know and don't know" from Proctor (1996), which explores these dynamics of knowledge with respect to cancer; also see Jain 2013. On tobacco marketing strategies informing efforts to mislead publics on scientific knowledge about climate change, see Oreskes and Conway 2011; on how the efficacy of new drugs is often better understood than their side effects and on clinical trial design, see Petryna 2009; on contested knowledge defining who counts as a survivor of and setting restrictions on the visibility of nuclear disaster, see Petryna 2002; on imperceptible risk in chemical exposure, see Murphy 2006, among others.

3. Short 2017. See chapter 6 on putting fire on the land. Tripp 2020 and Nikolakis and Roberts 2020.

4. Bachelard 1964:6. Also see Petryna 2018.

5. Bachelard coined the term "phenomenotechnique" for this tricky observational domain. He first used this concept to analyze the beginnings of microphysics (Castelão-Lawless 1995). On phenomenotechnique as "part thing, part theorem," see Rheinberger 2010:27. On the phenomenotechnique designating phenomena that are "constituted by the material setting of the laboratory," see Latour and Woolgar 1979:64.

On environmental disaster and abrupt climate change as new phenomena that eclipse these material settings, see Petryna 2013 and 2015.

6. Yedinak et al. 2018. On their assessment of the state of wildland fire-spread modeling, see Cohen and Finney 2010.

7. The connection was never lost if one takes Indigenous operating systems of fire into account (see chapter 6).

8. See Bankoff, Lubken, and Sand 2012:5. By 1835, in Rhode Island, fire-resistant factories were being built. Insurers, realizing a good rate-of-return, gave the owners credit for a good fire rating.

9. Albini 1976:6. Such an approach was preferred "even if the model overpredicts or underpredicts systematically, whether due to model inapplicability, model inaccuracy, or data errors."

10. Norman Maclean's *Young Men and Fire* (1992) cited in Andrews 2006.

11. The first quote is from Rothermel 1972:ii; the second is from Cohen's.

12. Bachelard 1964:6.

13. Cohen and Strohmaier 2020. "Defensible space" principles are contained in Cohen's Home Ignition Zone concept. He was also a codeveloper of the US National Fire Danger Rating System, which estimates fire danger for a given area. His research on fire behaviors, live fuels, ember ignitions, and structure ignitability has reshaped understandings of fire behavior.

14. Twelve firefighters died as they failed to outrun the fire; two helitack crew members on top of the ridge also perished. See Butler et al. 1998. This fire is the subject of John N. Maclean's book *Fire on the Mountain: The True Story of the South Canyon Fire* (1999).

15. On the limits of intuitive expertise for US wildfire-management programs, see Calkin and Mentis 2015.

16. Cohen 2000:1. The term wildland-urban interface is a misnomer because it assumes a fixed geographic component to the combustion patterns Cohen observed. Also see Pyne 2016:58.

17. See https://www.fs.usda.gov/about-agency/meet-forest-service.

18. This presumed efficiency can be interpreted as fueling the "disaster capitalism" (Klein 2007) that inhibits a "reality response."

19. The child's embodied response speaks to what anthropologist Marcel Mauss famously called "the techniques of the body" (1934). Following the "politics of things" (Bennett 2009) to the new materialisms (Coole and Frost 2010), dissonance as a form of acting in the world will resonate with Cohen's further insights.

20. Touring Paradise, California, after the devastating Camp Fire, former US president Trump sought to justify logging on forested public lands by suggesting that they should be "raked." Removing the very trees that are vital to slowing the release of carbon into the atmosphere creates more inoperable extremes down the road.

21. Ingalsbee 2020:174–175; 2017. Ingalsbee is executive director of the Firefighters United for Safety, Ethics, and Ecology (FUSEE).

22. Bill Tripp@CulturalFire, December 2. 2020. Also see Tripp 2020. On US wildfire fighters becoming stewards in restoration efforts, see Ingalsbee 2020.

23. He was referring to the Rothermel model: "everything that I was doing didn't fit with the Rothermel model." After his second experimental burn, he said, "I never made a Rothermel model calculation."

24. The federal government had been cutting funding for fire science; he left the Forest Service, worked for a heat exchanger manufacturer, and returned to the Riverside Fire Lab in 1988.

25. On fundamental discoveries about the mechanisms of wildfire spread, see Finney et al. 2015. Also see Tullis 2013.

Chapter 9. Going through the Porthole

1. Safety zones are a staple concept in US wildland fire suppression. Calculations of safety zone size have been based on radiative heat transfer; newer guidelines attempt to incorporate convective heat transfer.

2. "An entrapment may or may not include deployment of a fire shelter for its intended purpose. These situations may or may not result in injury. They include 'near misses.'" The National Wildfire Coordinating Group, https://www.nwcg.gov/term/glossary/entrapment.

3. "Wildland Fire Shelter: History and Development of the New Generation Fire Shelter" (undated).

4. Petrilli's experience and that of his surviving colleagues is recounted in Maclean 1999.

5. National Interagency Fire Center 2014.

6. National Interagency Fire Center 2014.

7. Ridler 2017.

8. Direct flame contact meant that temperatures inside fire shelters exceeded what they were designed to deflect. According to a formal investigation after Yarnell, "The fire shelter deployment was at a time and in a location where all fuels, weather, and topography aligned to produce rapid rates of spread, long flame lengths, flames being bent to nearly parallel with the surface." See Karels and Dudley 2013:79.

9. Wildland fires "are typically 1,600 degrees Fahrenheit; in some instances, temperatures can reach 2,000 degrees Fahrenheit. The most extreme temperature measured on a wildland fire was 2,400 degrees Fahrenheit." National Wildfire Coordinating Group, Fire Shelter Subcommittee, "Frequently Asked Questions about Fire Shelters," May 14, 2019, https://www.nwcg.gov/sites/default/files/memos/eb-m-19-001b.pdf.

10. Carroll 2013.

11. Webb and Gooden 2020. National differences in the use of the shelter can also reflect varying fuels and fire behaviors, operational requirements, and degrees of risk tolerance across countries.

12. van Stralen and Provansal 2007.

13. Chung 2013.

14. "Fire entities in Israel, Spain, Portugal and Cyprus all carry the U.S. Forest Service spec fire shelters. China, Chile and other European countries are now considering fire shelters." Petrilli, interviewed in Dotson's "Who Studies Fire Shelters? This Guy" (2021).

Chapter 10. Beneath the Airshow

1. Smith 2016. A notion of the "big lie," referring to efforts to call electoral results into question, permeated the 2020 presidential election. In wildfire discourse, it refers to unacknowledged safety issues; it has also been used to call out institutions for placing blame on employees for accidents that they did not have the means to deal with. See Ingalsbee 2010: xvi.

2. This title is in effect when a wildland firefighter works for a federal agency: On supporting safety, the National Wildfire Coordinating Group publishes guidelines, *The Incident Response Pocket Guide*, carried by firefighters in the field, and focusing on environmental factors that a firefighter can sense such as fuel moisture and temperatures, and observe, such as wind and atmospheric instability.

3. This bill was introduced on May 23, 2019, in a previous session of Congress, but it did not come up for a vote. "S. 1682—116th Congress: Wildland Firefighter Recognition Act," 2019, https://www.govtrack.us/congress/bills/116/s1682. O'Brien 2019.

4. Developed by the US Forest Service, the Work Capacity Test stresses the firefighter to assess aerobic endurance and muscle strength, as well as the ability to carry out strenuous labor for a prolonged time period.

5. As compared to a national suicide rate of 0.01 percent, the suicide rate for wildland firefighters in the US Forest Service is 0.3 percent, constituting a mental health crisis. See Hansman 2017; Gabbert 2017.

6. United States Forest Service and National Park Service 2016:11. On organizational responses to firefighter deaths and an "illusion of self-determinacy," see Desmond 2007.

7. Thermally driven (or diurnal) winds are pushed by cycles of heating and cooling at lower levels of terrain.

8. The information officer said there was no safety zone. The fire burned in the Bitterroot National Forest south of Hamilton, a local county seat. Erickson 2016.

9. In 2020, over two million acres burned in California; over a half a million burned in Oregon. Gollner 2020.

10. Less wind, cooler temperatures, and soaking rain helped.

11. Backus 2016.

12. On this point, also see Ingalsbee 2020.

13. Cohen, personal communication. Here, different cover types can emerge; species that tolerate low light levels might invade a vegetation type that is shade-intolerant. Add persistent drought on top of this shift and, instead of there being forest after the next fire, an invasive brush or grass appears and the forest disappears.

Horizon Work in a Time of Runaway Climate Change

1. Manabe 1998.

BIBLIOGRAPHY

Abatzoglou, John T., and A. Park Williams. 2016. "Impact of Anthropogenic Climate Change on Wildfire across Western U.S. Forests." *Proceedings of the National Academy of Sciences* 113(42):11770–11775.

Agee, James K., and Carl N. Skinner. 2005. "Basic Principles of Forest Fuel Reduction Treatments." *Forest Ecology and Management* 211:83–96.

Ahmann, Chloe, and Alison Kenner. 2020. "Breathing Late Industrialism." *Engaging Science, Technology, and Society* 6:416–438.

Albini, Frank. 1976. "Estimating Wildfire Behavior and Effects." USDA. Forest Service General Technical Report INT-30. https://www.fs.fed.us/rm/pubs_int/int_gtr030.pdf.

Andrews, Patricia. 2006. "Frank Albini, 1936–2005." *International Journal of Wildland Fire* 15:1–2. https://www.fs.fed.us/rm/pubs_other/rmrs_2006_andrews_p001.pdf.

Appadurai, Arjun. 2021. "The Scarcity of Social Futures in the Digital Era." In *Futures*, edited by Sandra Kemp and Jenny Andersson, 280–295. London: Oxford University Press.

Arendt, Hannah. 1961. *Between Past and Future*. New York: Penguin Books.

Ault, Alicia. 2016. "Can Animals Predict Earthquakes?" Smithsonianmag.Com, https://www.smithsonianmag.com/smithsonian-institution/ask-smithsonian-can-animals-predict-earthquakes-180960079/.

Azevedo, Tasso, and Nigel Sizer. 2019. "Brazil Has the Skill to Stop the Amazon Fires. It Lacks the Will." *Newsweek*, September 18. https://www.newsweek.com/amazon-fires-brazil-has-will-lacks-will-1460003.

Bachelard, Gaston. 1964. *The Psychoanalysis of Fire*. Boston: Beacon Press.

Backus, Perry. 2016. "Cooler Temperatures, Predicted Rain Could Help Firefighters Gain Control of Observation Blaze near Hamilton." *Missoulian*, July 5. http://missoulian.com/news/local/cooler-temperatures-predicted-rain-could-help-firefighters-gain-control-of/article_9c2b2399-2e22-5851-84a0-d0b011b9f34e.html.

Ballestero, Andrea. 2019. *A Future History of Water*. Durham, NC: Duke University Press.

Bankoff, Greg, Uwe Lubken, and Jordan Sand, eds. 2012. *Flammable Cities: Urban Conflagration and the Making of the Modern World*. Madison: University of Wisconsin Press.

Barash, Paul G., Bruce F. Cullen, Robert K. Stoelting, Michael Cahalan, M. Christine Stock, and Rafael Ortega, eds. 2019. *Clinical Anesthesia*. 7th ed. New York: Wolters Kluwer/Lippincott Williams & Wilkins.

Barreto, P., R. Pereira, A. Brandão, and S. Baima. 2017. "Will Meat-Packing Plants Help Halt Deforestation in the Amazon?" Imazon. https://imazon.org.br/en /publicacoes/will-meat-packing-plants-help-halt-deforestation-in-the-amazon/.

Barrett, Scott, Timothy M. Lenton, Antony Millner, Alessandro Tavoni, Stephen Carpenter, John M. Anderies, F. Stuart Chapin, Anne-Sophie Crépin, Gretchen Daily, Paul Ehrlich, Carl Folke, Victor Galaz, Terry Hughes, Nils Kautsky, Eric F. Lambin, Rosamond Naylor, Karine Nyborg, Stephen Polasky, Marten Scheffer, James Wilen, Anastasios Xepapadeas, and Aart de Zeeuw. 2014. "Climate Engineering Reconsidered." *Nature Climate Change* 4:527–529.

Basso, Keith. 1996. *Wisdom Sits in Places: Landscape and Language among the Western Apache*. Albuquerque: University of New Mexico Press.

Belval, E. J., C. S. Stonesifer, and D. E. Calkin. 2020. "Fire Suppression Resource Scarcity: Current Metrics and Future Performance Indicators." *Forests* 11(2):217. https://doi.org/10.3390/f11020217.

Bennett, Jane. 2010. *Vibrant Matter: A Political Ecology of Things*. Durham, NC: Duke University Press.

Berwyn, Bob. 2020. "Unchecked Global Warming Could Collapse Whole Ecosystems, Maybe Within 10 Years." *Inside Climate News*, April 8. https://insideclimatenews .org/news/07042020/global-warming-ecosystem-biodiversity-rising-heat-species ?utm_source=twitter&utm_medium=social.

Bigart, Robert, and Joseph McDonald. 2020. *"You Seem to Like Your Money, and We Like Our Country": A Documentary History of the Salish, Pend d'Oreille, and Kootenai Indians, 1875–1889*. Lincoln: University of Nebraska Press.

Biggs, Reinette, Stephen R. Carpenter, and William A. Brock. 2009. "Spurious Certainty: How Ignoring Measurement Error and Environmental Heterogeneity May Contribute to Environmental Controversies." *BioScience* (59)1: 65–76.

Bond, Horatio. 1946. *Fire and the Air War*. National Fire Protection Association.

Bonilla, Yarimar, and Marisol LeBrón, eds. 2019. *Aftershocks of Disaster: Puerto Rico before and after the Storm*. New York: Haymarket.

Boyd, Ian L. 2012. "The Art of Ecological Modeling." *Science* 337(6092):306–307.

Boyer, Dominic. 2019. *Energopolitics: Wind and Power in the Anthropocene* (single-authored volume of a collaborative duograph with C. Howe). Durham: Duke University Press.

Brechner, Verne L., Robert O. Bauer, Robert F. Wolff, Robert W. M. Bethune, and John B. Dillon. 1965. "Tests of Potassium Superoxide Canisters in a Small Fallout Shelter." *Public Health Reports*. 80(3):225–232.

Britannica Guide to Climate Change. 2008. Chicago: Encyclopaedia Britannica. http://eb.pdn.ipublishcentral.com/product/britannica-guide-to-climate-change.

Brogan, T.V.F. 2012. "Anthimeria." In *The Princeton Encyclopedia of Poetry and Poetics*. 4th ed., edited by Stephen Cushman, Clare Cavanaugh, Jahan Ramazani, and Paul Rouzer, 468–469. Princeton, NJ: Princeton University Press.

Brysse, Keynyn, Naomi Oreskes, Jessica O'Reilly, and Michael Oppenheimer. 2013. "Climate Change Prediction: Erring on the Side of Least Drama?" *Global Environmental Change* 23(1):327–337.

Butler, Bret W., Roberta A. Bartlette, Larry S. Bradshaw, Jack D. Cohen, Patricia L. Andrews, Ted Putnam, and Richard J. Mangan. 1998. "Fire Behavior Associated with the 1994 South Canyon Fire on Storm King Mountain, Colorado." USDA Forest Service Research Paper RMRS-RP-9. https://www.fs.fed.us/rm/pubs/rmrs_rp009.pdf.

Butler, Corey, Suzanne Marsh, Joseph W. Domitrovich, and Jim Helmkamp. 2017. "Wildland Firefighter Deaths in the United States: A Comparison of Existing Surveillance Systems." *Journal of Occupational and Environmental Hygiene* 14(4):258–270.

Cahill, Abigail E., Matthew E. Aiello-Lammens, M. Caitlin Fisher-Reid, Xia Hua, Caitlin J. Karanewsky, Hae Yeong Ryu, Gena C. Sbeglia, Fabrizio Spagnolo, John B. Waldron, Omar Warsi, and John J. Wiens. 2013. "How Does Climate Change Cause Extinction?" *Proceedings of the Royal Society B* 280:20121890. http://doi.org/10.1098/rspb.2012.1890.

Calkin, David E., and Mike Mentis. 2015. "Opinion: The Use of Natural Hazard Modeling for Decision Making under Uncertainty." *Forest Ecosystems* 2:11.

Callison, Candis. 2014. *How Climate Change Comes to Matter: The Communal Life of Facts.* Durham, NC: Duke University Press.

———. 2020. "The Twelve-Year Warning." *Isis* 111(1):129–137.

Cannon, Walter Bradford. 1915. *Bodily Changes in Pain, Hunger, Fear and Rage: An Account of Recent Researches into the Function of Emotional Excitement.* New York: D. Appleton and Company.

———. 1942. "'Voodoo' Death." *American Anthropologist* 44:169–181.

Canon, Gabrielle. 2020. "'I got the bug': A Pioneering Wildfire Fighter on the Thrills and Threats of the Job." *Guardian.* September 25.

Carney, Mark. 2015. "Breaking the Tragedy of the Horizon: Climate Change and Financial Stability" (speech given at Lloyd's of London), September 29.

Carpenter, Stephen R. 2003. *Regime Shifts in Lake Ecosystems: Pattern and Variation.* International Ecology Institute, Oldendorf/Luhe.

Carpenter, Stephen R., et al. 2011. "Early Warnings of Regime Shifts: A Whole-Ecosystem Experiment." *Science* 332:1079–1082.

Carr, E. Summerson. 2010. "Enactments of Expertise." *Annual Review of Anthropology* 39(1):17–32.

Carroll, Lucy. 2013. "Personal Fire Shelters Not Used in Australia." *Sydney Morning Herald,* July 2. http://www.smh.com.au/national/personal-fire-shelters-not-used-in-australia-20130702-2p9m2.html.

Castelão-Lawless, Teresa. 1995. "Phenomenotechnique in Historical Perspective: Its Origins and Implications for Philosophy of Science." *Philosophy of Science* 62 (1):44–59.

Chasing Ice (documentary film). 2013. Directed by Jeff Orlowski. Docurama Films.

Chung, Emily. 2013. "Why Emergency Fire Shelters Aren't Used in Canada." CBC. https://www.cbc.ca/news/technology/why-emergency-fire-shelters-aren-t-used-in-canada-1.1319366.

Ciavatta, Estevão. 2019. *Amazonia Undercover* (documentary film). Pindorama Filmes.

Clark, G. F., J. S. Stark, E. L. Johnston, J. W. Runcie, P. M Goldsworthy, B. Raymond, and M. J. Riddle. 2013. "Light-Driven Tipping Points in Polar Ecosystems." *Global Change Biology* 19(12):3749–3761.

Clark, Sara A., Andrew Miller, and Don L. Hankins—for the Karuk Tribe. 2020. *Good Fire: Current Barriers to the Expansion of Cultural Burning and Prescribed Fire in California and Recommended Solutions.* https://karuktribeclimatechangeprojects .com/good-fire/.

Climate and Traditional Knowledges Workgroup. 2014. "Guidelines for Considering Traditional Knowledges in Climate Change Initiatives." http://climatetkw .wordpress.com/.

Cohen, Jack D. 2000. "Examination of the Home Destruction in Los Alamos Associated with the Cerro Grande Fire—July 10, 2000." USDA Forest Service, Rocky Mountain Research Station, Missoula, Montana.

Cohen, Jack D., and Mark A. Finney. 2010. "An Examination of Fuel Particle Heating during Fire Spread." *VI International Conference on Forest Fire Research*, edited by D. X. Viegas.

Cohen, Jack, and Dave Strohmaier. 2020. "Community Destruction during Extreme Wildfires Is a Home Ignition Problem." *Wildfire Today*, September 21. https:// wildfiretoday.com/2020/09/21/community-destruction-during-extreme -wildfires-is-a-home-ignition-problem/.

Coleman, W. 1964. *Georges Cuvier, Zoologist: A Study in the History of Evolution Theory.* Cambridge, MA: Harvard University Press.

Confederated Salish and Kootenai Tribes. 2005. *Beaver Steals Fire: A Salish Coyote Story.* Lincoln, NE: Bison Books.

Confederated Salish and Kootenai Tribes. 2006. *Beaver Steals Fire / Fire on the Land: A 2-DVD Educational Set.* Pablo, MT: Confederated Salish and Kootenai Tribes.

Confederated Salish and Kootenai Tribes of the Flathead Reservation. 2013. *Climate Change Strategic Plan.* http://csktclimate.org/downloads/Climate%20Change%20 Strategic%20Plan/CSKT%20Climate%20Change%20Adaptation%20Plan%204 .14.16.pdf.

Coole, Diana H., and Samantha Frost, eds. 2010. *New Materialisms Ontology, Agency, and Politics.* Durham, NC: Duke University Press.

Cox, Alicia. 2017. "Settler Colonialism." Oxford Bibilographies. https://www.oxford bibliographies.com/view/document/obo-9780190221911/obo-9780190221911 -0029.xml.

Crapanzano, Vincent. 2003. *Imaginative Horizons: An Essay in Literary-Philosophical Anthropology.* Chicago: University of Chicago Press.

Crew, Bec. 2017. "This Small Lake in Africa Once Killed 1,700 People Overnight, and We Still Don't Know Why." *Science Alert*, April 15.

Cronon, William. 1995. "The Trouble with Wilderness; or, Getting Back to the Wrong Nature." In *Uncommon Ground: Toward Reinventing Nature*, edited by W. Cronon, 69–90. New York: Norton.

Curwen, Thomas. 2017 "California's Deadliest Wildfires Were Decades in the Making. 'We have forgotten what we need to do to prevent it.'" *Los Angeles Times*, October 22.

Davis, Heather, and Zoe Todd. 2017. "On the Importance of a Date, or, Decolonizing the Anthropocene." *ACME: An International Journal for Critical Geographies* 16(4):761–780.

Davis, Steven J., Long Cao, Ken Caldeira, and Martin I. Hoffert. 2013. "Rethinking Wedges." *Environmental Research Letters* 8(1). doi: 10.1088/1748–9326/8/1/011001.

Dejours, Pierre. 1978. "Carbon Dioxide in Water- and Air-Breathers." *Respiration Physiology* 33(1): 121–128.

de la Cadena, Marisol. 2015. "Uncommoning Nature." *E-flux Journal* 65:1–8.

DeLanda, M. 2010. *Deleuze: History and Science*. New York: Atropos.

Desmond, Matthew. 2007. *On the Fireline: Living and Dying with Wildland Firefighters*. Chicago: University of Chicago Press.

Diamond, Jared. 2011. *Collapse: How Societies Choose to Fail or Succeed*. New York: Penguin Books.

Diamond, Sarah E. 2018. "Contemporary Climate-Driven Range Shifts: Putting Evolution Back on the Table." *Functional Ecology* 32:1652–1665.

Dotson, Travis. 2021. "Who Studies Fire Shelters? This Guy." Wildland Fire Lessons Learned Center. March 23.

Downey, Clare. 2017. "Why Are California's Wildfires So Out of Control?" (video). *Guardian*, December 8. https://www.theguardian.com/world/video/2017/dec/08/why-are-californias-wildfires-so-out-of-control-video-explainer.

Dumit. Joseph. 2014. "Writing the Implosion: Teaching the World One Thing at a Time." *Cultural Anthropology* 29(2):344–362.

Dunning, Philo, et al. 1884. "Two Hundred Tons of Dead Fish, Mostly Perch, at Lake Mendota, Wisconsin." *Bulletin of the United States Fish Commission*, 439–443.

Ebert, C.H.V. 1963. "Hamburg's Fire Storm Weather." *NFPA Quarterly* 56:253–260.

Eddy, Melissa, Jack Ewing, Megan Specia, and Steven Erlanger. 2021. "European Floods Are Latest Sign of a Global Warming Crisis." *New York Times*, July 18.

Eden, Lynn. 2006. *Whole World on Fire: Organizations, Knowledge, and Nuclear Weapons Devastation*. Ithaca, NY: Cornell University Press.

Edsall, Thomas B. 2015. "Whose Neighborhood Is It?" *New York Times*, September 9.

Egan, Timothy. 2009. *The Big Burn: Teddy Roosevelt and the Fire That Saved America*. Boston: Mariner Books.

Einhorn, Catrin, Maria Magdalena Arréllaga, Blacki Migliozzi, and Scott Reinhard. 2020. "The World's Largest Tropical Wetland Has Become an Inferno." *New York Times*, October 13. https://www.nytimes.com/interactive/2020/10/13/climate/pantanal-brazil-fires.html.

Epler, G. R. 1989. "Silo-Filler's Disease: A New Perspective." *Mayo Clinical Proceedings* 64:368–370.

Erickson, David. 2016. "Observation Fire 15 Percent Contained, New Fire Breaks Out near Lincoln." *Missoulian*, July 4.

Erickson, Doug. 2019. "The Ho-Chunk's Ancestral Home." OnWisconsin. https://onwisconsin.uwalumni.com/on_campus/the-ho-chunks-ancestral-home/.

Eshel, Amir. 2012. *Futurity: Contemporary Literature and the Quest for the Past*. Chicago: University of Chicago Press.

Fagundes, Guilherme Moura. 2019. "Fire Normativities: Environmental Conservation and Quilombola Forms of Life in the Brazilian Savanna." *Vibrant: Virtual Brazilian Anthropology* 16. http://www.vibrant.org.br/lastest-issue-v-16-2019/.

Ferguson, Brodie, Júlia Sekula, and Ilona Szabó. 2020. "Technology Solutions for Supply Chain Traceability in the Brazilian Amazon: Opportunities for the Financial Sector." Igarapé Institute. https://igarape.org.br/wp-content/uploads/2020/08/2020-08-24-AE-48_Amazonia-Technology-Solutions.pdf.

Finney, Mark A., Jack D. Cohen, Jason M. Forthofer, Sara S. McAllister, Michael J. Gollner, Daniel J. Gorham, Kozo Saito, Nelson K. Akafuah, Brittany A. Adam, and Justin D. English. 2015. "Role of Buoyant Flame Dynamics in Wildfire Spread." *Proceedings of the National Academy of Sciences* 112(32):9833–9838.

Finney, M. A., J. D. Cohen, S. S. McAllister, and W. M. Jolly. 2013. "On the Need for a Theory of Wildland Fire Spread." *International Journal of Wildland Fire* 22(1):25–36.

Finney, M. A., I. C. Grenfell, and C. W. McHugh. 2009. "Modeling Containment of Large Wildfires Using Generalized Linear Mixed-Model Analysis." *Forest Science* 55:249–255.

Fischer, Michael M. J. 2009. *Anthropological Futures*. Durham, NC: Duke University Press.

———. 2018. *Anthropology in the Meantime: Experimental Ethnography, Theory, and Method for the Twenty-First Century*. Durham, NC: Duke University Press.

Flock, Elizabeth, and Joshua Barajas. 2018. "They Reported Sexual Harassment. Then the Retaliation Began." March 1. https://www.pbs.org/newshour/nation/they-reported-sexual-harassment-then-the-retaliation-began.

Forbes, Stephen A. 1887. "The Lake as a Microcosm." *Bulletin of the Peoria Scientific Association*, 77–87. http://people.wku.edu/charles.smith/biogeog/FORB1887.htm.

Forthofer, Jason, and Scott Goodrick. 2011. "Review of Vortices in Wildland Fire." *Journal of Combustion* 2011. https://www.hindawi.com/journals/jc/2011/984363/#B10.

Fortun, Kim. 2001. *Advocacy after Bhopal: Environmentalism, Disaster, New Global Orders*. Chicago: University of Chicago Press.

Fourth National Climate Assessment (NCA4). 2018. "Indigenous Peoples Terminology for the Fourth National Climate Assessment." http://www7.nau.edu/itep/main/tcc/docs/resources/Indigenous%20Peoples%20Terminology%20for%20NCA4_final.pdf.

Fowler, Cynthia T., and James R Welch. 2018. *Fire Otherwise: Ethnobiology of Burning for a Changing World*. Salt Lake City: The University of Utah Press.

Friedrich, Jorg. 2006. *The Fire: The Bombing of Germany, 1940–1945*. Edited by Allison Brown. New York: Columbia University Press.

Gabbert, Bill. 2017. "Suicide Rate among Wildland Firefighters Is 'Astronomical.'" *Wildfire Today*, November 4.

———. 2021. "Bill Introduced to Raise the Max-Out Limit on Overtime Pay for Federal Employees." *Wildfire Today*, January 29.

Gadamer, Hans-Georg. 1997. *Truth and Method*. 2nd ed. New York: The Crossroad Publishing Corporation.

Ghosh, Amitav. 2017. *The Great Derangement: Climate Change and the Unthinkable*. Chicago: University of Chicago Press.

Gifford, R. 2011. "The Dragons of Inaction: Psychological Barriers That Limit Climate Change Mitigation and Adaptation." *American Psychologist* 66(4):290–302.

Gilpin, Emilee. 2019. "Urgency in Climate Change Advocacy Is Backfiring, Says Citizen Potawatomi Nation Scientist." *National Observer*, February 15.

Gladwell, Malcolm. 2002. *The Tipping Point: How Little Things Can Make a Big Difference*. New York: Back Bay Books.

Gladwin, Thomas. 1970. *East Is a Big Bird: Navigation and Logic on Puluwat Atoll*. Cambridge, MA: Harvard University Press.

Goeman, Mishuana. 2008. "From Place to Territories and Back Again: Centering Storied Land in the Discussion of Indigenous Nation-Building." *International Journal of Critical Indigenous Studies* 1(1):34. https://doi.org/10.5204/ijcis.v1i1.20.

Gollner, Michael. 2020. "How Do We Solve the 'Wicked Problem'? The State of Fire Science in California." California Fire Science Seminar Series. https://frg.berkeley.edu/how-do-we-solve-the-wicked-problem-the-state-of-fire-science-in-california/.

Günel, Gökçe. 2021. "Leapfrogging to Solar." *South Atlantic Quarterly* 120 (1):163–175.

Gottbrath, Laurin-Whitney. "California Law to Ease Process for Former Inmates to Become Professional Firefighters." *Axios*, September 11. https://www.axios.com/newsom-california-inmate-firefighters-wildfires-5254e522-3781-48c1-a439-26281ad12053.html.

Gould, Stephen Jay. 1971. "D'Arcy Thompson and the Science of Form." *New Literary History* 2(2):229–258.

Grandin, Temple, and Gary C. Smith. 2004. "Animal Welfare and Humane Slaughter." http://www.grandin.com/references/humane.slaughter.html.

Green, Miranda. 2020. "Native Americans Feel Double Pain of COVID and Fires 'Gobbling Up the Ground.'" *Kaiser Health News*, October 5.

Greendeer, Kendra 2019. "The Land Remembers Native Histories." Edge Effects, November 21. https://edgeeffects.net/native-histories/.

Griffen, Blaine D., and John M. Drake. 2009. "Scaling Rules for the Final Decline to Extinction." *Proceedings of the Royal Society* 276(1660):1361–1367.

Grinberg, Emanuella, Eliott C. McLaughlin, and Christina Zdanowicz. 2018. "A Northern California Fire Is Growing at a Rate of about 80 Football Fields per Minute." CNN. November 11.

Guterl, Fred. 2012. *The Fate of the Species: Why the Human Race May Cause Its Own Extinction and How We Can Stop It*. New York: Bloomsbury.

Halbwachs, Michel, Jean-Christophe Sabroux, and Gaston Kayser. 2019. "Final Step of the 32-Year Lake Nyos Degassing Adventure: Natural CO_2 Recharge Is to Be Balanced by Discharge through the Degassing Pipes." *Journal of African Earth Sciences* 167.

Hansman, Heather. 2017. "A Quiet Rise in Wildland-Firefighter Suicides." *Atlantic*, October 29. https://www.theatlantic.com/health/archive/2017/10/wildland-firefighter-suicide/544298/.

Haraway, Donna J. 2016. *Staying with the Trouble: Making Kin in the Chthulucene*. Durham, NC: Duke University Press.

Harjo, Laura. 2019. *Spiral to the Stars: Mvskoke Tools of Futurity*. Tucson: University of Arizona Press.

Harwood, Tony. 2021. "Indigenous Fire Management: Returning Fire to the Land" (lecture). Swan Valley Connections, August 5. https://www.youtube.com/watch?v=1fuXEuNjNjY.

Hay, Andrew. 2019. "Let It Burn: U.S. Fights Wildfires with Fire." Reuters World News, August 28. https://www.reuters.com/article/us-usa-wildfires-management-idUSKCN1VI19G.

Hayhurst, E. R., and E. Scott. 1914. "Four Cases of Sudden Death in a Silo." *Journal of the American Medical Association* 63:1570–1572.

Holcomb, R. C. 1916. "Atmosphere." *United States Naval Medical Bulletin* 10:430–465.

Holling, C. S. 1973. "Resilience and Stability of Ecological Systems." *Annual Review of Ecology and Systematics* 4:1–23.

———. 1998. "Two Cultures of Ecology." *Conservation Ecology* 2(2). https://www.ecologyandsociety.org/vol2/iss2/art4/.

Holthaus, Eric. 2017 "The First Wintertime Megafire in California History Is Here." Grist, December 8. https://grist.org/article/the-first-wintertime-megafire-in-california-history-is-here/.

Howe, Cymene. 2019. *Ecologics: Wind and Power in the Anthropocene* (single-authored volume of a collaborative duograph with D. Boyer). Durham: Duke University Press.

Hughes, Terry P., et al. 2012. "Living Dangerously on Borrowed Time during Slow, Unrecognized Regime Shifts." *Trends in Ecology and Evolution* 28(3):149–155.

Hughes, Terry P., Kristen D. Anderson, Sean R. Connolly, Scott F. Heron, James T. Kerry, Janice M. Lough, Andrew H. Baird, Julia K. Baum, Michael L. Berumen, Tom C. Bridge, Danielle C. Claar, C. Mark Eakin, James P. Gilmour, Nicholas A. J. Graham, Hugo Harrison, Jean-Paul A. Hobbs, Andrew S. Hoey, Mia Hoogenboom, Ryan J. Lowe, Malcolm T. Mcculloch, John M. Pandolfi, Morgan Pratchett, Verena Schoepf, Gergely Torda, and Shaun K. Wilson. 2018. "Spatial and Temporal Patterns of Mass Bleaching of Corals in the Anthropocene." *Science* 359(6371):80–83.

Huth, John. 2013. *The Lost Art of Finding Our Way*. Cambridge, MA: Belknap Press of Harvard University Press.

Hwang, In Cheol, et al. 2013. "Clinical Changes in Terminally Ill Cancer Patients and Death within 48 H: When Should We Refer Patients to a Separate Room?" *Support Care Cancer* 21:835–840.

Ingalsbee, Timothy. 2010. "Foreword." In *Inferno by Committee: A History of the Cerro Grande (Los Alamos) Fire, America's Worst Prescribed Fire Disaster*, xv–xxi. Bloomington, IN: Trafford Publishing.

———. 2014. "Pyroganda: Creating New Terms and Identities for Promoting Fire Use in Ecological Fire Management." In *Proceedings: Wildland Fire in the Appalachians: Discussions among Managers and Scientists*, edited by Thomas A. Waldrop, 170–177. Asheville, NC: USDA Forest Service, Southern Research Station.

———. 2017. "Whither the Paradigm Shift? Large Wildland Fires and the Wildfire Paradox Offer Opportunities for a New Paradigm of Ecological Fire Management." *International Journal of Wildland Fire* 26(7):557–561. https://doi.org/10.1071/WF17062.

Ingold, Tim. 2011. *Being Alive: Essays on Movement, Knowledge and Description*. New York: Routledge.

International Energy Agency. 2021. *Global Energy Review 2021*, IEA, Paris. https://www.iea.org/reports/global-energy-review-2021.

Irfan, Umair. 2020. "Why We're More Confident Than Ever That Climate Change Is Driving Disasters." *Vox*, September 30.

Ishimatsu, A., M. Hayashi, K.-S. Lee, T. Kikkawa, and J. Kita. 2005. "Physiological Effects on Fishes in a High-CO_2 World." *Journal of Geophysical Research* 110:C09S09, doi:10.1029/2004JC002564.

Jain, S. Lochlann. 2013. *Malignant: How Cancer Becomes Us*. Berkeley: University of California Press.

James, John T., and Ariel Macatangay. N.d. "Carbon Dioxide—Our Common 'Enemy.'" Houston, Texas: NASA/Johnson Space Center. https://ntrs.nasa.gov/archive/nasa/casi.ntrs.nasa.gov/20090029352.pdf.

Jamison, M. 2005. "Researchers Look to Native History to Gather Ideas about Forests and Flames." https://www.frames.gov/catalog/43234.

Jasanoff, Sheila. 2016. *The Ethics of Invention: Technology and the Human Future*. New York: Norton.

Jauss, Hans Robert. 1982. "Literary History as a Challenge to Literary Theory." in *Toward an Aesthetic of Reception*, trans. Timothy Bahti, 21–22. Minneapolis: University of Minnesota Press.

Jolly, W. Matt, Mark A. Cochrane, Patrick H. Freeborn, Zachary A. Holden, Timothy J. Brown, Grant J. Williamson, and David M.J.S. Bowman. 2015 "Climate-Induced Variations in Global Wildfire Danger from 1979 to 2013." *Nature Communications* 6. https://doi.org/10.1038/ncomms8537.

Kahneman, Daniel, and Gary Klein. 2009. "Conditions for Intuitive Expertise: A Failure to Disagree." *American Psychologist* 64(6):515–526.

Karels, Jim, and Mike Dudley. 2013. *Yarnell Hill Serious Accident Investigation Report*. Arizona State Forestry Division.

Karnauskas, K. B., S. L. Miller, and A. C. Schapiro. 2020. "Fossil Fuel Combustion Is Driving Indoor CO_2 Toward Levels Harmful to Human Cognition." *GeoHealth*, 4, e2019GH000237. https://doi.org/10.1029/2019GH000237.

Kimmerer, Robin Wall. 2000. "Native Knowledge for Native Ecosystems." *Journal of Forestry* 98(8):4–9. https://doi.org/10.1093/jof/98.8.4.

———. 2013. *Braiding Sweetgrass: Indigenous Wisdom, Scientific Knowledge and the Teachings of Plants*. Minneapolis, MN: Milkweed Editions.

———. 2021. "The Serviceberry: An Economy of Abundance." https://emergence magazine.org/story/the-serviceberry/.

King, Matthew Wilburn. 2019. "How Brain Biases Prevent Climate Action." BBC, March 7. https://www.bbc.com/future/article/20190304-human-evolution -means-we-can-tackle-climate-change.

Klein, Naomi. 2007. *The Shock Doctrine: The Rise of Disaster Capitalism*. New York: Picador.

Knowles, Scott. 2011. *The Disaster Experts: Mastering Risk in Modern America*. Philadelphia: University of Pennsylvania Press.

Kolbert, Elizabeth. 2021. *Under a White Sky: The Nature of the Future*. New York: Crown.

Kopenawa, Davi. 2013. *The Falling Sky: Words of a Yanomami Shaman*. Cambridge, MA: Belknap Press of Harvard University Press.

Kosek, Jake. 2006. *Understories: The Political Life of Forests in Northern New Mexico*. Durham, NC: Duke University Press.

Krenak, Ailton. 2020. *Ideas to Postpone the End of the World*. Toronto: House of Anansi Press.

Kusakabe, M., T. Ohba, I. Issa, Y. Yoshida, H. Satake, T. Ohizumi, W. C. Evans, G. Tanyileke, and G. W. Kling. 2008. "Pre- and Post-Degassing Evolution of CO_2 in Lakes Monoun and Nyos, Cameroon." *Geochemical Journal* 42:93–118.

Lahsen, Myanna. 2009. "A Science-Policy Interface in the Global South: The Politics of Carbon Sinks and Science in Brazil." *Climate Change* 97:339–372.

Lake, Frank Kanawha. 2021. "Indigenous Fire Stewardship: Federal/Tribal Partnerships for Wildland Fire Research and Management." *Fire Management Today* 79(1):30–39. https://www.fs.fed.us/psw/publications/lake/psw_2021_lake001.pdf.

Lakoff, Andrew. 2017. *Unprepared: Global Health in a Time of Emergency*. Berkeley: University of California Press.

Langlois, Krista. 2017. "What It's Like Being a Woman in the Male-Dominated World of Wildland Firefighting." Outside. https://www.outsideonline.com/2089206/what -its-being-woman-male-dominated-world-wildland-firefighting.

Latour, Bruno, and Steve Woolgar. 1979. *Laboratory Life: The Construction of Scientific Facts*. Princeton, NJ: Princeton University Press.

Lavender, George. 2013. "Fighting Fires Is Big Business for Private Companies." *Earth Island Journal*, October 29.

Lenton, Timothy M. 2011. "Early Warning of Climate Tipping Points." *Nature Climate Change* 1:201–209.

Lenton, Timothy M., Hermann Held, Elmar Kriegler, Jim W. Hall, Wolfgang Lucht, Stefan Rahmstorf, and Hans Joachim Schellnhuber. 2008. "Tipping Elements in the Earth's Climate System." *PNAS* 105(6):1786–1793.

Lewis, David. 1971. "'Expanding' the Target in Indigenous Navigation." *Journal of Pacific History* 6:83–95.

Liebmann, Matthew J., Joshua Farella, Christopher I. Roos, Adam Stack, Sarah Martini, and Thomas W. Swetnam. 2016. "Native American Depopulation, Reforestation, and Fire Regimes in the Southwest United States, 1492–1900 CE." *PNAS* 113(6):E696–E704. https://doi.org/10.1073/pnas.1521744113.

Lifton, Robert Jay. 2017. "Malignant Normality." *Dissent.* https://www.dissentmagazine.org/article/malignant-normality-doctors.

Lin, Brenda B., and Brian Petersen. 2013. "Resilience, Regime Shifts, and Guided Transition under Climate Change: Examining the Practical Difficulties of Managing Continually Changing Systems." *Ecology and Society* (18)1. https://doi.org/10.5751/ES-05128-180128.

Lindqvist, Sven. 2000. *A History of Bombing.* New York: The New Press.

Lovejoy, Thomas E., and Carlos Nobre. 2018. "Amazon Tipping Point." *Science Advances* 5(12).

LTER Network. 1979. "Long-Term Ecological Research Concept and Measurement Needs." https://lternet.edu/documents/1979-report-on-long-term-ecological-research-concept-and-measurement-needs/.

Lubchenco, Jane, and Jack Hayes. 2012. "A Better Eye on the Storm." *Scientific American* 306:68–73.

Lucas, K. A., J. M. Orient, A. Robinson, H. Maccabee, P. Morris, G. Looney, and M. Klinghoffer. 1990. "Efficacy of Bomb Shelters: With Lessons from the Hamburg Firestorm." *Southern Medical Journal* 83(7):812–20.

Maclean, John N. 1999. *Fire on the Mountain: The True Story of the South Canyon Fire.* New York: Harper Perennial.

Maclean, Norman. 1992. *Young Men and Fire.* Chicago: University of Chicago Press.

Magnuson, J. J. 1990. "Long-Term Ecological Research and the Invisible Present." *BioScience* 40(7): 495–501.

———. 1995. "The Invisible Present." In *Ecological Time Series,* edited by T. M. Powell and J. H. Steele, 448–464. Boston: Springer.

Magnuson, J. J., C. J. Bowser, and A. L. Beckel. 1983. "The Invisible Present: Long-Term Ecological Research on Lakes." *Letters & Science Magazine* (University of Wisconsin–Madison), fall, 3–6.

Mahony, Martin, and Mike Hulme. 2012. "The Color of Risk: An Exploration of the IPCC's 'Burning Embers' Diagram." *Spontaneous Generations* 6(1):75–89.

Manabe, Syukuro. 1998. Oral History Interview by Paul Edwards. American Institute of Physics. https://www.aip.org/history-programs/niels-bohr-library/oral-histories/32158-2.

Mason, L., G. White, G. Morishima, E. Alvarado, L. Andrew, F. Clark, M. Durglo, J. Durglo, J. Eneas, J. Erickson, et al. 2012. "Listening and Learning from Traditional Knowledge and Western Science: A Dialogue on Contemporary Challenges of Forest Health and Wildfire." *Journal of Forestry* 110(4):187–193. https://doi.org/10.5849/jof.11-006.

Mauss, Marcel. 1973. "Techniques of the Body." *Economy and Society* 2(1):70–88.

Mayr, E. 1982. *The Growth of Biological Thought: Diversity, Evolution, and Inheritance.* Cambridge, MA: Harvard University Press.

Mbembe, Achille. 2020. "The Universal Right to Breathe." Translated by Carolyn Shread. *Critical Inquiry* 47(52). https://www.journals.uchicago.edu/doi/full/10.1086/711437.

McClintock, Anne. 2020. "Monster: A Fugue in Fire and Ice." *E-Flux Journal.* https://www.e-flux.com/architecture/oceans/331865/monster-a-fugue-in-fire-and-ice/.

Migliozzi, Blacki, Scott Reinhard, Nadja Popovich, Tim Wallace, and Allison McCann. 2020. "Record Wildfires on the West Coast Are Capping a Disastrous Decade." *New York Times*, September 24. https://www.nytimes.com/interactive/2020/09/24/climate/fires-worst-year-california-oregon-washington.html.

Montana Public Radio. 2021. Fireline podcast, "Episode 4: The Gift of Fire." (Justin Angle, Victor Yvellez, Nick Mott). March 30. https://www.firelinepodcast.org/episode-4-the-gift-of-fire/.

Mooney, Chris, 2020. "Two Major Antarctic Glaciers Are Tearing Loose from Their Restraints, Scientists Say." *Washington Post*, September 14. https://www.washingtonpost.com/climate-environment/2020/09/14/glaciers-breaking-antarctica-pine-island-thwaites/.

Moore, Amelia. 2016. "Anthropocene Anthropology: Reconceptualizing Contemporary Global Change." *Journal of the Royal Anthropological Institute* 22(1):27–46.

Morton, Timothy. 2013 *Hyperobjects: Philosophy and Ecology after the End of the World.* Minneapolis: University of Minnesota Press.

Moyers, Bill. 2017. "The Dangerous Case of Donald Trump: Robert Jay Lifton and Bill Moyers on 'A Duty to Warn.'" https://billmoyers.com/story/dangerous-case-donald-trump-robert-jay-lifton-bill-moyers-duty-warn/.

Muñoz, José Esteban. 2009. *Cruising Utopia: The Then and There of Queer Futurity.* New York: New York University Press.

Murphy, Michelle. 2006. *Sick Building Syndrome and the Problem of Uncertainty: Environmental Politics, Technoscience, and Women Workers.* Durham, NC: Duke University Press.

Napier, David. 2014. "The New Sociobiology: Symbiosis and Local Meaning." Paper presented at the "End of Biodeterminism?" Conference at the Centre for Cultural Epidemics, Aarhus, Denmark, October 1.

NASA. 2020. "World of Change: Antarctic Sea Ice." NASA Earth Observatory, February 18. https://earthobservatory.nasa.gov/world-of-change/LarsenB.

National Geographic. 2019. "Climate Milestone: Earth's CO_2 Level Passes 400 ppm." March 29. https://www.nationalgeographic.org/article/climate-milestone-earths-co2-level-passes-400-ppm/#:~:text=On%20May%209%2C%202013%2C%20an,million%20years%20of%20Earth%20history.

National Research Council. 2013. *Abrupt Impacts of Climate Change: Anticipating Surprises.* Washington, DC: National Academies Press.

National Wildfire Coordinating Group. 2014. "Wildland Fire Safety Training Annual Refresher: 1994 South Canyon Fire on Storm King Mountain" (video). https://www.nwcg.gov/publications/training-courses/rt-130/case-studies/cs201.

Neale, Timothy. 2020. "What Are Whitefellas Talking About When We Talk About "Cultural Burning"?" Inside Story. April 17. https://insidestory.org.au/what-are-whitefellas-talking-about-when-we-talk-about-cultural-burning/.

Neale, Timothy, Rodney Carter, Rodney Nelson, and Mick Bourke. 2019. "Walking Together: A Decolonising Experiment in Bushfire Management on Dja Dja Wurrung Country." *Cultural Geographies* 26(3):341–359.

Newburger, Emma. 2019. "Massive Arctic Wildfires Emitted More CO_2 in June Than Sweden Does in an Entire Year." CNBC, August 19.

Nietzsche, Friedrich. 1909. "The Use and Abuse of History." In *Thoughts Out of Season: Part II*, translated by Adrian Collins. Edinburgh: T. N. Foulis.

———. 1997. *Nietzsche: Untimely Meditations*. Translated by R. J. Hollingdale. Edited by Daniel Breazeale. Cambridge: Cambridge University Press. First published in 1874.

Nikolakis, William D., and Emma Roberts. 2020. "Indigenous Fire Management: A Conceptual Model from Literature." *Ecology and Society* 25. https://www.ecologyandsociety.org/vol25/iss4/art11/.

Nixon, Rob. 2011. *Slow Violence and the Environmentalism of the Poor*. Cambridge, MA: Harvard University Press.

———. 2020. "All Tomorrow's Warnings." Public Books, August 13. https://www.publicbooks.org/all-tomorrows-warnings/.

NOAA. 2021. "Billion-Dollar Weather and Climate Disasters: Time Series." https://www.ncdc.noaa.gov/billions/time-series.

Nolan, Janne E. 2004. "Measuring the Unthinkable." *Science* 303(5665):1772–1773.

Nordenson, Guy. 2016. "Probabilistic Coastal Hazards Mapping for the United States." Princeton Environmental Institute seminar, Princeton, NJ, September 20.

Norgaard, Kari Marie. 2011. *Living in Denial: Climate Change, Emotions, and Everyday Life*. Cambridge, MA: MIT Press.

———. 2019. *Salmon and Acorns Feed Our People*. New Brunswick, NJ: Rutgers University Press.

O'Brien, Edward. 2019. "'Firefighter' or 'Forestry Tech' and Why It Matters." Montana Public Radio, May 17. https://www.mtpr.org/post/firefighter-or-forestry-tech-and-why-it-matters.

O'Brien, Jean M. 2010. *Firsting and Lasting: Writing Indians Out of Existence in New England*. Minneapolis: University of Minnesota Press.

O'Nell, Theresa D. 1996. *Disciplined Hearts: History, Identity, and Depression in an American Indian Community*. Berkeley: University of California Press.

Oreskes, Naomi, and Erik Conway. 2011. *Merchants of Doubt: How a Handful of Scientists Obscured the Truth on Issues from Tobacco Smoke to Global Warming*. New York: Bloomsbury.

Oxfam Media Briefing. 2020. "Confronting Carbon Inequality: Putting Climate Justice at the Heart of the COVID-19 Recovery." September 21.

Pacala, Stephen, and Robert Socolow. 2004. "Stabilization Wedges: Solving the Climate Problem for the Next 50 Years with Current Technology." *Science* 305(5686):968–972.

Parunak, H.V.D., T. C. Belding, and S. A. Brueckner. 2008. "Prediction Horizons in Agent Models." In *Engineering Environment-Mediated Multiagent Systems* (proceedings of Satellite Conference at ECCS 2007, Dresden, Germany), edited by Danny Weyns, Sven A. Brueckner, and Yves Demazeau, 88–102. Springer.

Peterson, Karen. 2017. "Smokejumpers Drop Down on Local Field." *Valley Journal*, May 17.

Petryna, Adriana. 2002. *Life Exposed: Biological Citizens after Chernobyl*. Princeton, NJ: Princeton University Press.

———. 2009. *When Experiments Travel: Clinical Trials and the Global Search for Human Subjects*. Princeton, NJ: Princeton University Press.

———. 2013. "On the Origins of Extinction." *Limn*, no. 3, 50–53.

———. 2015. "What Is a Horizon? Navigating Thresholds in Climate Change Uncertainty." In *Modes of Uncertainty: Anthropological Cases*, edited by P. Rabinow and L. Samimian-Darash, 147–164. Chicago: University of Chicago Press.

———. 2018. "Wildfires at the Edges of Science: Horizoning Work amid Runaway Change." *Cultural Anthropology* 339(4):570–595.

Petryna, Adriana, and Paul Mitchell. 2017. "On the Nature of Catastrophic Forms." *BioSocieties* 2(3):343–366.

Pimm, Stuart L. 2009. "Climate Disruption and Biodiversity." *Current Biology* 19 (14):R595–R601.

Pizer, William A. 2017. "What's the Damage from Climate Change?" *Science* 356(6345):1330–1331. http://science.sciencemag.org/content/356/6345/1330.full.

Plucinski, M. P. 2019. "Contain and Control: Wildfire Suppression Effectiveness at Incidents and across Landscapes." *Current Forestry Reports* 5(1):20–40. https://doi.org/10.1007/s40725-019-00085-4.

Povoledo, Elisabetta. 2017. "Can Animals Predict Earthquakes? Italian Farm Acts as a Lab to Find Out." *New York Times*, June 17.

Proctor, Robert. 1995. *The Cancer Wars: How Politics Shapes What We Know and Don't Know about Cancer*. New York: Basic Books.

Pryor, A. J., and C. H. Yuill. 1966. *Mass Fire Life Hazard*. San Antonio, TX: Southwest Research Institute. https://www.frames.gov/catalog/11783.

Puerini, James, and Gerald Torres. 2020. "Op-Ed: Don't Just Cheer Wildland Firefighters as Heroes. Give Them Affordable Healthcare." *LA Times*, June 20. https://www.latimes.com/opinion/story/2020-06-20/wildfires-health-insurance-covid-19-firefighters.

Pyne, Stephen J. 1982. *Fire in America: A Cultural History of Wildland and Rural Fire*. Princeton, NJ: Princeton University Press.

———. 2001. *Year of the Fires: Story of the Great Fires of 1910*. New York: Viking Press.

———. 2016. *The Northern Rockies: A Fire Survey*. Tucson: University of Arizona Press.

———. 2021. *The Pyrocene: How We Created an Age of Fire, and What Happens Next*. Berkeley: University of California Press.

Randalls, Samuel. 2015 "Joshua P. Howe. Behind the Curve: Science and the Politics of Global Warming" (book review). *Isis* 106(2):503–505. https://doi.org/10.1086/682823.

Rempfer, Kyle. 2018. "Could the Air Force Bomb Wildfires into Submission?" *Air Force Times*, August 10. https://www.airforcetimes.com/news/your-air-force/2018/08/10/could-the-air-force-bomb-wildfires-into-submission/.

Reo, N. J., and A. K. Parker. 2014. "Re-thinking Colonialism to Prepare for the Impacts of Rapid Environmental Change." *Climatic Change*120:671–683.

Report on Long-Term Ecological Research Concept and Measurement Needs. 1979. Summary of a Workshop, Institute of Ecology, Indianapolis, Indiana, June 25–27. https://lternet.edu/wp-content/uploads/2010/12/79Workshop.pdf.

Rheinberger, Hans-Jorg. 2010. *An Epistemology of the Concrete: Twentieth-Century Histories of Life*. Durham, NC: Duke University Press.

Ribe, Tom. 2010. *Inferno by Committee: A History of the Cerro Grande (Los Alamos) Fire, America's Worst Prescribed Fire Disaster*. Bloomington, IN: Trafford Publishing.

Richards, O. 1955. "D'Arcy W. Thompson's Mathematical Transformation and the Analysis of Growth." *Annals of the New York Academy of Sciences* 63(4):456–473.

Ridler, Keith. 2017. "US Plan to Improve Firefighter Shelters Falters." Associated Press, May 4. https://apnews.com/a3a037d922c240a0bafa7406b9623702/APNewsBreak:-US-plan-to-improve-firefighter-shelters-falters.

Rifkin, Mark. 2017. *Beyond Settler Time: Temporal Sovereignty and Indigenous Self-Determination*. Durham, NC: Duke University Press.

Ritchie, Hannah. 2019. "Who Has Contributed Most to Global CO_2 Emissions?" https://ourworldindata.org/contributed-most-global-co2.

Rocha, J. C., G. Peterson, Ö Bodin, and S. Levin S. 2018. "Cascading Regime Shifts within and across Scales." *Science* 362(6421):1379–1383.

Rockström, Johan, et al. 2009. "Planetary Boundaries: Exploring the Safe Operating Space for Humanity." *Ecology and Society* 14(2):32.

Rojas, David. 2016. "Climate Politics in the Anthropocene and Environmentalism beyond Nature and Culture in Brazilian Amazonia." *Political and Legal Anthropology Review* 39(1):16–32. https://doi.org/10.1111/plar.12128.

Rothermel, Richard. 1972. *A Mathematical Model for Predicting Fire Spread in Wildland Fuels*. Res. Pap. INT-115. Ogden, UT: US Department of Agriculture, Intermountain Forest and Range Experiment Station.

Salish-Pend d'Oreille Culture Committee and Elders Cultural Advisory Council, Confederated Salish & Kootenai Tribes. 2008. *The Swan Valley Massacre: A Brief History*. http://www.salishaudio.org/documents.

———. 2019. *The Salish People and the Lewis and Clark Expedition*. Rev. ed. Lincoln: University of Nebraska Press.

Samimian-Darash, Limor. and Paul Rabinow, eds. 2015. *Modes of Uncertainty: Anthropological Cases*. Chicago: University of Chicago Press.

Sapinski, J. P., Holly Jean Buck, and Andreas Malm, eds. 2020. *Has It Come to This? The Promises and Perils of Geoengineering on the Brink*. New Brunswick, NJ: Rutgers University Press.

Scheffer, Marten. 2009. *Critical Transitions in Nature and Society*. Princeton, NJ: Princeton University Press.

Scheffer, M., S. Carpenter, J. A. Foley, C. Folke, and B. Walke. 2001. "Catastrophic Shifts in Ecosystems." *Nature* 413:591–596.

Scheffer, Marten, et al. 2012. "Anticipating Critical Transitions." *Science* 338:334–348.

Sebald, W. G. 2011. "Air War and Literature." In *On the Natural History of Destruction*. New York: Modern Library.

Séliš-Q̓lispe Culture Committee. 2019a. "Skʷskʷstúlexʷ | Names upon the Land: The Salish-Kalispel Ethnogeography Project." http://www.csktsalish.org/index.php /ethnogeography/ethnogeography-booklet.

Séliš-Q̓lispe Culture Committee. 2019b. "Salish-Pend d'Oreille Placename Signs on Highway 93." http://www.csktsalish.org/index.php/audio/salish-pend-d-oreille -placename-signs-on-u-s-highway-93.

Shapin, Steven, and Simon Schaffer. 1985. *Leviathan and the Air-Pump: Hobbes, Boyle, and the Experimental Life*. Princeton, NJ: Princeton University Press.

Sharpe, Christina. 2016. *In the Wake: On Blackness and Being*. Durham, NC: Duke University Press.

Short, Karen C. 2017 "Rethinking Performance Measurement in U.S. Federal Wildland Fire Management." Presentation slide deck, Fire Behavior Workshop, Northern Rockies Training Center, Missoula, Montana, February 22–24.

Shukman, David. 2013. "Carbon Dioxide Passes Symbolic Mark." BBC. May 10.

Singer, Emily. 2014. "The Remarkable Self-Organization of Ants." *Guardian*, April 11. https://www.theguardian.com/science/2014/apr/11/ants-self-organization -quanta.

Smith, Mark. 2016. "The Big Lie." Wildland Fire Leadership. http://wildlandfireleader-ship.blogspot.com/2016/06/the-big-lie.html.

Socolow, R. 2011. "Wedges Reaffirmed," Climate Central, September 27. https://www .climatecentral.org/blogs/wedges-reaffirmed.

———. 2020. "Witnessing for the Middle to Depolarize the Climate Change Conversation." *Daedalus* 149(4):46–66.

Solomon, Susan, Gian-Kasper Plattner, Reto Knutti, and Pierre Friedlingstein. 2008. "Irreversible Climate Change Due to Carbon Dioxide Emissions." *PNAS* 106(6):1704–1709.

Spence, Katy. 2017. "Putting Fire on the Land: Montana Tribes Use Traditional Knowledge to Restore Forests." Treesource, July 24. https://treesource.org/news /management-and-policy/native-american-fire-use/.

Squyres, Steven. *Roving Mars: "Spirit," "Opportunity," and the Exploration of the Red Planet*. New York: Hyperion, 2005.

Steffen, Will, Johan Rockström, Katherine Richardson, Timothy M. Lenton, Carl Folke, Diana Liverman, Colin P. Summerhayes, Anthony D. Barnosky, Sarah E. Cornell, Michel Crucifix, Jonathan F. Donges, Ingo Fetzer, Steven J. Lade, Marten Scheffer, Ricarda Winkelmann, and Hans Joachim Schellnhuber. 2018. "Trajectories of the Earth System in the Anthropocene." *PNAS* 115 (33):8252–8259.

Sternberg, Esther M. 2002. "Walter B. Cannon and 'Voodoo' Death: A Perspective from 60 Years On." *American Journal of Public Health* 92(10):1564–1566.

Sullivan, A. L. 2018. "A Review of Wildland Fire Spread Modelling, 1990–Present." October 23. https://arxiv.org/pdf/0706.3074.pdf.

Swaney, Ron. 2021. "Wildland Fire Update." 44th Annual National Indian Timber Symposium. June 7.

Swetnam, Thomas W., Joshua Farella, Christopher I. Roos, Matthew J. Liebmann, Donald A. Falk, and Craig D. Allen. 2016. "Multiscale Perspectives of Fire, Climate and Humans in Western North America and the Jemez Mountains, USA." *Philosophical Transactions of the Royal Society B* 371(1696):20150168.

Taylor, Arnold H. 2011. *The Dance of Air and Sea: How Oceans, Weather, and Life Link Together.* Oxford: Oxford University Press.

Telch, Michael J., David Rosenfield, Han-Joo Lee, et al. 2012. "Emotional Reactivity to a Single Inhalation of 35% Carbon Dioxide and Its Association with Later Symptoms of Posttraumatic Stress Disorder and Anxiety in Soldiers Deployed to Iraq." *Archives of General Psychiatry* 69(11):1161–1168. doi:10.1001/archgenpsychiatry.2012.8.

Thom, René. 1975. *Structural Stability and Morphogenesis: An Outline of a General Theory of Models.* Reading: W. A. Benjamin.

Thomas, Jordan. 2020. "A Note from the Fireline: Climate Change and the Colonial Legacy of Fire Suppression." The Drift. https://www.thedriftmag.com/a-note-from-the-fireline/.

Thomas, Martin. 2004. *The Artificial Horizon: Reading a Colonised Landscape.* Melbourne: Melbourne University Publishing.

Thompson, D. W. 1942. *On Growth and Form.* 2nd ed. Cambridge: Cambridge University Press.

Thompson, Matthew P., Donald G. MacGregor, Christopher J. Dunn, David E. Calkin, and John Phipps. 2018. "Rethinking the Wildland Fire Management System." *Journal of Forestry* 116(4):382–390. https://doi.org/10.1093/jofore/fvy020.

Todd, Zoe. 2014. "Fish Pluralities: Human-Animal Relations and Sites of Engagement in Paulatuuq, Arctic Canada." *Etudes/Inuit/Studies* 38(1–2):217–238.

———. 2018. "Refracting the State through Human-Fish Relations: Fishing, Indigenous Legal Orders and Colonialism in North/Western Canada." *Decolonization: Indigeneity, Education & Society* 7(1):60–75.

Tripp, Bill. 2020. "Our Land Was Taken. But We Still Hold the Knowledge of How to Stop Mega-Fires." *Guardian*, September 16. https://www.theguardian.com/commentisfree/2020/sep/16/california-wildfires-cultural-burns-indigenous-people.

Troisi, F. M. 1957. "Delayed Death Caused by Gassing in a Silo Containing Green Forage." *British Journal of Industrial Medicine* 14:56–58.

Tsing, Anna Lowenhaupt. 2012. "On Nonscalability: The Living World Is Not Amenable to Precision-Nested Scales." *Common Knowledge* 18(3):505–524.

———. 2015. *The Mushroom at the End of the World: On the Possibility of Life in Capitalist Ruins.* Princeton, NJ: Princeton University Press.

Tuck, Eve, and K. Wayne Yang. 2012. "Decolonization Is Not a Metaphor." *Decolonization: Indigeneity, Education & Society* 1 (1):1–40.

Tullis, Paul. 2013. "Into the Wildfire." *New York Times*, September 19. http://www
.nytimes.com/2013/09/22/magazine/into-the-wildfire.html?pagewanted=all.

United States Forest Service and National Park Service. 2016 "Strawberry Fire Fatality:
Learning Review Report." Wildland Fire Lessons Learned Center.

van Stralen, Daved, and Gary Provansal. 2007. "The French Connection." https://irp
-cdn.multiscreensite.com/ba5e96f3/files/uploaded/Foreign_Fire.pdf.

Verma, Vandi, John Langford, and Reid Simmons. 2001. "Non-Parametric Fault Identification for Space Rovers." http://www.ri.cmu.edu/pub_files/pub2/verma_vandi
_2001_1/verma_vandi_2001_1.pdf.

Verran, Helen. 2002. "A Postcolonial Moment in Science Studies: Alternative Firing
Regimes of Environmental Scientists and Aboriginal Landowners." *Social Studies
of Science* 32(5–6):729–762.

Vertesi, Janet. 2015. *Seeing Like a Rover: How Robots, Teams, and Images Craft Knowledge of Mars*. Chicago: University of Chicago Press.

Vettese, Troy, and Drew Pendergrass. 2020. "Our Global Fire Crisis Is the Sign of a
Dying Biosphere. But We Can Take Action." *Guardian*, December 8.

Vizenor, Gerald. 2008. "Aesthetics of Survivance." In *Survivance: Narratives of Native
Presence*, edited by Gerald Vizenor, 1–23. Lincoln: University of Nebraska Press.

Waddington, C. H. 1942. "Canalization of Development and the Inheritance of
Acquired Characters." *Nature* 150(3811):563–565.

———. 1957. *The Strategy of the Genes*. London: George Allen & Unwin.

Wallace-Wells, David. 2019a. "Time to Panic." *New York Times*, February 16.

———. 2019b. *The Uninhabitable Earth: Life after Warming*. New York: Duggan.

Wang, Rong, et al. 2012. "Flickering Gives Early Warning Signals of a Critical Transition to a Eutrophic Lake State." *Nature* 492:419–422.

Watt-Cloutier, Sheila. 2020. "Upirngasaq (Arctic Spring)." *Granta*. https://granta.com
/upirngasaq-arctic-spring/.

Watts, Vanessa. 2013. "Indigenous Place-Thought and Agency amongst Humans and
Non Humans (First Woman and Sky Woman Go on a European World Tour!)."
Decolonization: Indigeneity, Education & Society 2 (1):20–34.

Webb, Andrew, and Andy Gooden. 2020. "Engineering a Safer Crew Protection System." International Association of Wildland Fire.

Weber, Jeremy. 2017. "Preserving the Past, Protecting the Future." *Lake County Leader*,
October 19.

Weheliye, Alexander G. 2014. *Habeas Viscus: Racializing Assemblages, Biopolitics, and
Black Feminist Theories of the Human*. Durham, NC: Duke University Press.

White, Germaine. 2007. "The Gift of Fire." Wildland Fire Lessons Learned Center.

White, James. 2014. Public briefing on National Research Council report, "Abrupt
Impacts of Climate Change: Anticipating Surprises." Washington, DC, January 10.
https://www.youtube.com/watch?v=uh3auNaQbhc.

Whitington, Jerome. 2020. "Earth's Data: Climate Change, Thai Carbon Markets, and
the Planetary Atmosphere. American Anthropologist, 122(4): 814–826.

Whyte, Kyle Powys. 2017. "Our Ancestors' Dystopia Now: Indigenous Conservation and the Anthropocene." In *The Routledge Companion to the Environmental Humanities*, edited by Ursula K. Heise, Jon Christensen, and Michelle Niemann, 206–215. New York: Routledge.

———. 2018a. "Indigenous Climate Justice: Intersecting Science, Politics and Education" (lecture). International Symposium on Indigenous Communities and Climate Change, December 7. Princeton University.

———. 2018b. "Indigeneity in Geoengineering Discourses: Some Considerations." *Ethics, Policy & Environment* 21(3): 289–307.

Wildcat, Daniel R. 2009. *Red Alert! Saving the Planet with Indigenous Knowledge.* Golden, CO: Fulcrum Publishing.

Wolchover, Natalie. 2019. "A World without Clouds." Quanta Magazine, February 25. https://www.quantamagazine.org/cloud-loss-could-add-8-degrees-to-global -warming-20190225/.

Wynter, Sylvia. 2003. "Unsettling the Coloniality of Being/Power/Truth/Freedom: Towards the Human, after Man, Its Overrepresentation—An Argument." *CR: The New Centennial Review* 3(3):257–337.

Yamauchi, H., H. Uchiyama, N. Ohtani, and M. Ohta. 2014. "Unusual Animal Behavior Preceding the 2011 Earthquake off the Pacific Coast of Tohoku, Japan: A Way to Predict the Approach of Large Earthquakes." *Animals* 4(2):131–145.

Yazzie, Victoria Lynn. 2006. "A Cultural Ethic in Tribal Forest Management and Self-Determination: The Human Dimension of Silviculture." PhD diss., University of Montana.

Yedinak, Kara M., Eva K. Strand, J. Kevin Hiers, and J. Morgan Varner. 2018. "Embracing Complexity to Advance the Science of Wildland Fire Behavior." *Fire* 1(2):20.

Zagorski, Nick. 2005. "Profile of Stephen R. Carpenter." *PNAS* 102(29):9999–10001.

Zahara, Alex. 2020. "Breathing Fire into Landscapes That Burn: Wildfire Management in a Time of Alterlife." *Engaging Science, Technology, and Society* 6:555–585.

Zee, Jerry. 2017. "Holding Patterns: Sand and Political Time at China's Desert Shores." *Cultural Anthropology* (32)2:215–241.

INDEX

Page references followed by *fig* indicate an illustrated figure or photograph.

illusions of control over, 151–52; stabilization triangle concept on, 34–37*fig*. *See also* CO_2 emissions

Guató Silva, Sandra, 24–25

Hamburg bombing (World War II), 13, 14*fig*

Hamilton's rule, 162n.4

Hansen, James, 155

Haraway, Donna, 72

Harjo, Laura, 28, 81

Harwood, Tony, 90, 93, 94, 170n.40

heat-resistant firefighter uniform, 138*fig*

Ho-Chunk forced removal (1837), 22

Holcomb, R. C., 7

Holling, C. S. (Buzz), 43

Home Ignition Zone concept, 115–16, 173n.13

Homo sapiens, 6–7

horizon-deprived: description of, 94–95; replace extended horizon thinking for, 150; "tragedy of the horizon" of, 154

horizon work/horizoning: applied to system environments, 48; description and purpose of, 3, 5; expanding horizons to estimate ecological shifts, 25–29; expanding the target in navigation through, 166n.6; exploring climate futures in terms of, 150–51; the fifty-year horizon tool for, 38; framing how wildland firefighters think, 95; future-indifferent sense horizon-deprivation converse of, 94–95; Hamilton's rule defining individual selfishness of, 162n.4; original meaning as points of reference, 47; personal dimension of, 152–53; philosophy, literary, and social sciences on, 163n.14; scale shifts and, 49, 51, 58, 60; as scaling tool in an invisible present, 48, 49; spurious certainty creating short-term, 41–42; as supplied to wildland firefighter routines, 95; as wayfinding tool for a durable world, 152

hotshot wildland firefighter crews, 97–99

human responsiveness asset, 154–55

Hurricane Sandy (Superstorm Sandy), 38

Hutton, James, 52

hypercapnia, 17

"hyperobjects," 74

hypothetical universes, 108, 113, 121

Incident Command System (ICS), 98

The Incident Pocket Guide, 175n.2

incremental emissions notion, 26

Indian Self-Determination and Education Assistance Act (1975), 82

Indigenous people (CSKT): Big Burn (1910) burning land of, 68, 89, 90; camas-gathering site maps (1940s) by, 93, 94; climate change impact on frontline communities of, 168–69n.8; Confederated Salish and Kootenai Tribes (CSKT), 83, 87, 90, 170n.37; continuous occupancy of western Montana by, 169n.15; fire as luminous instrument and, 86; first contact with Lewis and Clark expedition (1805), 170n.3; Flathead Reservation, 88, 90, 91, 93, 169n.10; horizon thinking evidenced by, 79, 91–92; land management and political self-determination of, 80–90, 170n.32, 171n.44; place-names of the, 83–84, 169–70n.20; population decline (1492–1900 CE) among, 168n.3; safeguarding lifeworlds principles of, 28; Salish-Kootenai Fire History Project (2005), 82–83, 84, 85, 169nn.13, 20; settler colonialism form of domination of, 168n.4; settler time vs. stabilization of Indigenous time, 22, 80–81, 84–87, 163n.10; stewardship of agential fish, 163n.8; Swan Valley Massacre (1908), 89; *Sxʷpaám* (Makes Fire or Fire Setter) role of, 81, 83; Tribal Self-Governance Act (1994), 82, 93

Indigenous wildland firefighters: Chippy Creek wildfire (2007) fought by, 92–93; elimination of fire suppression practices of, 80–81, 88–90; examining environmental options

A NOTE ON THE TYPE

This book has been composed in Adobe Text and Gotham.
Adobe Text, designed by Robert Slimbach for Adobe,
bridges the gap between fifteenth- and sixteenth-century
calligraphic and eighteenth-century Modern styles.
Gotham, inspired by New York street signs, was designed
by Tobias Frere-Jones for Hoefler & Co.